Calculus

Elliot C. Gootman

Professor of Mathematics
The University of Georgia

BARRON'S

Dedication

To Marilyn, Elissa, Jennifer, and Michael. Your enthusiasm helped me start this book, and your encouragement, patience, support, and love helped me finish it.

All inquiries should be addressed to:
Barron's Educational Series, Inc.
250 Wireless Boulevard
Hauppauge, New York 11788

Library of Congress Catalog No. 97-13991

International Standard Book No. 0-8120-9819-6

Library of Congress Cataloging-in-Publication Data
Gootman, Elliot C.
    Calculus / Elliot C. Gootman.
        p.   cm. — (College review series)
    Includes index.
    ISBN 0-8120-9819-6
    1. Calculus–Outlines, syllabi, etc.   I. Title.   II. Series.
QA303.G65   1997
515—dc21   97-13991
                                                  CIP

PRINTED IN THE UNITED STATES OF AMERICA
9 8 7

CONTENTS

Contents   **v**

11.4   Volumes of Solids of Revolution—The Method of
         Cylindrical Shells 275
11.5   Summary of Main Points 288
11.6   Exercises 290

12 • Motion **293**
12.1   Simple Initial Value Problems 293
12.2   Motion with Constant Acceleration 295
12.3   Summary of Main Points 300
12.4   Exercises 301

APPENDICES

A • The Trigonometric Functions **305**
A.1   Definitions 305
A.2   Identities 308
A.3   Derivatives 309
A.4   Antiderivatives 311

B • Exponential and Logarithmic Functions **315**
B.1   Definitions 315
B.2   Identities 316
B.3   Derivatives 317
B.4   Antiderivatives 319

• **Answers to Exercises** **322**

• **Glossary** **336**

• **Index** **339**

INTRODUCTION

This book is meant to be an accessible introduction to the main ideas, methods and applications of first year calculus. It is accessible because it assumes very little prior knowledge of mathematics. In particular, this book pays special attention to two common areas of weakness many students bring to their study of calculus—algebra and reading comprehension. Rather than adding yet more boring chapters of preliminary material, however, background weaknesses are addressed throughout the book, as the need arises. At the appropriate places in the book, the reader will find brief reviews of algebraic techniques and geometric formulas. In the chapter on word problems, hints are given on the issue of reading comprehension. The tone of the book is informal and conversational. Ideas and terminology are explained with words, with detailed, down-to-earth explanations in English, before they are explained with formulas. The meaning of all the symbolism is carefully explained, so that the formulas will make sense to the reader. This approach is designed to overcome the problem of math anxiety, which is for many students the root cause of their lack of success in beginning calculus.

It is my belief that to understand mathematics, one has to understand the *meanings* of the technical terms and symbols, and not just the rules for manipulating them.

As the explanations given are unhurried, and as the purpose of the book is to deal with the *main* ideas, a number of topics normally covered in a standard first year calculus text are not covered in this book. In addition, I have mostly avoided the unusual constructions or counterexamples that serve to illustrate the fine points of when a certain definition might hold or might fail. Rather, I try to stress throughout the "typical" example or application, which serves to illuminate or motivate the main concept or idea under discussion. I hope that for the reader who needs it, this book will be the beginning of the story of calculus. It is certainly not the end of the story. If you plan to become a structural engineer, helping to design bridges, tunnels, and buildings, then I trust you'll realize that you need to know more than you will find here. If, however, you are trying to "get into" calculus, and traditional textbooks are not being of much help, then I hope this book will be a useful guide for you.

If you are reading this introduction, then you probably want to learn more about calculus than you already know. Maybe you are the proverbial "intelligent layman" who just wants to get some idea of what calculus is all about. Maybe you had calculus many years ago and now find that you need a little bit of it for your work, but don't remember any. This book is for you.

But most of all, this book is for high school or college students who are currently taking a course in calculus, and not doing so well at it. Its primary purpose is as a supplement for a calculus student in trouble. If you cannot seem to jump into a standard calculus text and figure out what's going on, then I hope you can jump into this book and understand. If you are missing the forest for the trees, so that you are learning all sorts of techniques of computation, but have lost sight of what the whole course is all about, read this book. The detailed and simple English explanations will serve to remind you of the fundamentals of the subject, and, most important, will help you to connect the symbols and computations with a verbal description of what they mean. In addition, for the convenience of students, I have included an appendix containing a brief summary of the trigonometric, exponential, and logarithmic functions.

Each chapter of the book contains numerous worked examples. Read an example, and then try to do it on your own. If you get stuck, see how I have worked the example and try to get yourself unstuck. Then try to finish it yourself. Working on examples, doing computations, and solving problems are important in mathematics. You cannot really be sure you understand the ideas and concepts if you cannot implement them. Furthermore, the process of implementing the basic ideas and concepts actually enhances your understanding of these ideas. Try not to cheat by reading my solutions too quickly. Really give these examples a good effort of your own.

I know that many students do not like to read. You may become impatient with all of my words, and want to go straight to the worked examples. Many students just want to see how to work the problem. Let me remind you about the old fable of the tortoise and the hare. Slow and steady often wins the race. The right way to learn mathematics is first to try to understand, and then use your understanding to work problems. Often the right way, although it seems slower and less efficient at the time, is in the long run the fastest and most efficient way of reaching your goal.

Good luck with calculus. I hope you enjoy this book, and that it provides you with a real learning experience. If you need more than what this book offers, use what you have learned here as a base. Once you have really mastered the fundamental concepts, you will have a solid foundation for mastering additional, more technical and more complicated material later.

Elliot C. Gootman

About the Author

The author is a professional research mathematician, and also a professional teacher of mathematics. He has taught, for the past twenty-five years, a wide range of mathematics courses, from precalculus to research seminars for Ph.D. students. Calculus, however, has consistently been a staple of his teaching schedule. Students have commented, in anonymous evaluations, that "he explains things so clearly it is hard not to understand them," and "he could teach calculus to a cat."

1
EQUATIONS, FUNCTIONS, AND GRAPHS

1.1 EQUATIONS

One way in which humanity progresses in its understanding of the universe is by discovering relationships between various objects, concepts, quantities, and so on. Even when these relationships are fuzzy and somewhat unclear, they promote understanding by forcing us to consider more closely questions of cause and effect. Our natural impulse to try to improve our knowledge of these relationships, to make them clearer, more direct, and more quantified also leads to further progress. One example of current interest is the relationship between the amount of pollutants we pump into the atmosphere and the future average temperature of the earth. This relationship is, of course, at the heart of the controversy concerning the issue of global warming. If we perfectly understood the relationship, and had it completely quantified, there would be no controversy.

Clearly, your understanding of a relationship between two quantities is sharpest when this relationship can be completely quantified and expressed in an **equation**. This is why equations are so important and so widely studied—they are explicit expressions of relationships between quantities. Thus, $x^3 - 2xy + y^2 = 5$ is an equation relating x and y. A particular pair of values, one for x and the other for y, satisfies the equation if the equality is true when you substitute in the values. Thus the pair $x = 2$ and $y = 1$ satisfies the equation, because when you substitute in the value 2 for x and the value 1 for y, you get

$$2^3 - (2 \cdot 2 \cdot 1) + 1^2 = 8 - 4 + 1 = 5$$

Thus, the value of the left-hand side of the equation does indeed equal the value of the right-hand side of the equation. The pair $x = 3$ and $y = 2$ does not satisfy the equation, because when you substitute in the value 3 for x and 2 for y, you get

$$3^3 - (2 \cdot 3 \cdot 2) + 2^2 = 27 - 12 + 4 = 19$$

The value 19 of the left-hand side of the equation certainly does not equal the value 5 on the right-hand side of the equation.

Example 1.1 Does the pair of numbers $x = 2$ and $y = 3$ satisfy the equation $x^2 - 4xy = y^2 + 16$?

1

Solution: The technique is simply to substitute in the value 2 for x and 3 for y, and see whether or not the left-hand side equals the right-hand side. Substituting in on the left-hand side you get

$$2^2 - (4 \cdot 2 \cdot 3) = 4 - 24 = -20$$

Substituting in on the right-hand side gives

$$3^2 + 16 = 9 + 16 = 25$$

As the two sides are not equal, the pair $x = 2$ and $y = 3$ does not satisfy the equation.

Example 1.2 Find a pair of numbers, one for x and the other for y, which does satisfy the equation $x^2 - 4xy = y^2 + 16$.

Solution: Sometimes problems like this can be tricky. One possibility is just to pick any value for x, substitute it into the equation, so that you get an equation with only y in it, and then try to solve for y. A good value to try for x in problems like this is 0 (since that usually makes some terms disappear). If $x = 0$ then the whole left-hand side equals 0, and the equation becomes $0 = y^2 + 16$. This has no solution in y, because y^2 is always a nonnegative number, so that $y^2 + 16$ will always be at least 16, and thus can never be 0. Well, let's try substituting in the value $y = 0$ (another way of making some terms disappear). The terms $4xy$ and y^2 equal 0, so the equation reduces to $x^2 = 16$. There are two solutions, $x = 4$ and $x = -4$. Thus the pair $x = 4$ and $y = 0$ satisfies the equation. The pair $x = -4$ and $y = 0$ also satisfies the equation.

1.2 FUNCTIONS

Equations into Functions

There is an obvious difference between an equation like $x^3 - 2xy + y^2 = 5$ and an equation like $y = x^2 - 4x + 2$. In the first equation, the two variables x and y are intertwined. In the second equation, the variable y stands alone on the left side of the equal sign, while the expression to the right of the equal sign involves only the other variable, x. This second type of equation is closely related to the notion of a **function**. In this second equation,

$$y = x^2 - 4x + 2 \tag{1.1}$$

the expression on the right-hand side can be viewed as a recipe, formula, rule, or procedure for starting out with a given value of x and computing precisely one corresponding value of y. The rule is that you take the given number x, square it, subtract from the square the product of 4 times x, and then add 2. Such a rule or procedure is at the heart of the idea of a function. More specifically:

DEFINITION 1.1. *A function is a formula, rule, or procedure for taking a set of numbers and assigning to each number x in the set precisely one value.*

Often functions are represented by letters such as f or g. If f is a function, then $f(x)$ stands for the value that the function f assigns to the number x. It is also called "the value of f at x." The symbol $f(x)$ is read as "f of x." The letter x is thought of as the **independent variable**, because it can be picked to be more or less any number. The dependence of the values of f upon x, that is, the fact that a value $f(x)$ is determined by the formula, rule, or procedure from the value of x, is expressed by saying that "f **is a function of** x."

An expression such as

$$f(x) = x^2 - 4x + 2 \qquad (1.2)$$

is just an exact description of the formula for assigning values. Given a number x, then the value of f at x, $f(x)$, is found by squaring x, then subtracting the product of 4 times x, and then adding 2. The term x inside the parentheses in $f(x)$ is what you feed into the formula. It is very useful to think of a function as actually doing something to what you give it.

If f is a function of x, then $y = f(x)$ is a special kind of equation, in which one of the quantities, y, appears alone on the left side of the equal sign, and the expression on the right side of the equal sign involves only the other quantity, x. Since the values of y depend upon those of x, y is thought of as the **dependent variable**. Conversely, when you have this special kind of equation, such as $y = x^2 - 4x + 2$, it is common to think of the right-hand side as defining a function $f(x)$, and of the equation as being simply $y = f(x)$. In other words, the equation is simply thought of as *defining* y as a function of x.

Note to students:

1. Often, in mathematics, functions are defined more generally. The particular procedure for computing values does not have to be given explicitly. In fact, the variables and their assigned values do not even have to be numbers. This book will not deal with these more general types of functions.

2. The advantage of the function notation $f(x)$ is that the symbol $f(3)$ can be used to represent the value of the function when $x = 3$. If you just use the equation notation $y = x^2 - 4x + 2$, and want to talk about the value of y that satisfies the equation when $x = 3$, you must use all those words, or say something such as "when $x = 3, y = -1$."

Some equations are simple enough so that with a little work and algebra, you can actually rewrite the equation and solve for one of the quantities in terms of the other, to get one quantity basically expressed as a function of the other. In other equations, you may in principle be able to solve for one quantity in terms of another, but in practice the algebra might be too difficult. In still other equations, it is not even possible in principle to solve for one quantity in terms of the other.

Consider the equation $2x - y = 7$. This equation is simple enough so that with a very small amount of algebraic work, you can easily rewrite it to solve explicitly for one of the quantities in terms of the other. Thus $2x - y = 7$ is equivalent to $y + 7 = 2x$, which is equivalent to $y = 2x - 7$, or $y = f(x)$, with $f(x) = 2x - 7$. Now y is written explicitly as a function of x, with the expression $2x - 7$ on the right-hand side of the equal sign giving the rule for computing a value of y from a value of x: multiply the x value by 2, and then subtract 7. Thus $f(5) = 2 \cdot 5 - 7 = 10 - 7 = 3$.

Example 1.3 For the function $f(x) = 2x - 7$, compute $f(x^3), f(3x)$, and $f(-4)$.

Solution: Remember, the term inside the parentheses is what you input into the rule or procedure. The expression $2x - 7$ gives the procedure for what you do to the input: multiply it by 2, and then subtract 7. When the input is x^3, you multiply that by 2, and subtract 7, to get $f(x^3) = 2 \cdot x^3 - 7 = 2x^3 - 7$. When the input is $3x$, that is what you multiply by 2, and then subtract 7 from, to get $f(3x) = 2 \cdot 3x - 7 = 6x - 7$. When the input is -4, you multiply -4 by 2, and then subtract 7, to get $f(-4) = (2 \cdot -4) - 7 = -8 - 7 = -15$.

Look at the equation $2x - y = 7$ again. You may notice that you can just as easily solve for x in terms of y, so that $2x = y + 7$ and $x = \dfrac{y + 7}{2}$. In this form, you can view x as essentially being given as a function of y, for example as $x = g(y)$ with

$$g(y) = \frac{y + 7}{2}$$

Example 1.4 Rewrite the equation $x^2 - 2xy = 5y + 4$ so that y is expressed as a function of x.

Solution: The instructions basically mean to try to solve for y in terms of x. The goal is to isolate y on one side of the equation, and have no y terms at all on the other side of the equation. The first step is to collect all the y terms on one side, and all the terms without y on the other. Doing this step by step gives

$$\begin{aligned}
x^2 - 2xy &= 5y + 4 \\
-2xy &= -x^2 + 5y + 4 \\
-2xy - 5y &= -x^2 + 4 \\
(-2x - 5)y &= -x^2 + 4 \\
y = f(x) &= \frac{-x^2 + 4}{-2x - 5}
\end{aligned}$$

Example 1.5 Can you rewrite the equation $x^2 + y^2 = 16$ so that y is expressed as a function of x?

Solution: This example shows that you sometimes must exercise great care in trying to change a general equation into a function. You may recognize this equation as the equation of a circle, centered at the origin, of radius 4. If you try to isolate the y terms on one side, and all the terms without y on the other, you get $y^2 = 16 - x^2$. Now it looks as if you can solve for y simply by taking square roots of both sides to get

$$y = \sqrt{16 - x^2}$$

Suppose $x = 0$, so that $16 - x^2 = 16$. The square root of 16 is, by definition, a number whose square equals 16. There are really two such numbers, 4 and -4. You can easily see that both pairs $x = 0$, $y = 4$ and $x = 0$, $y = -4$ satisfy the equation $x^2 + y^2 = 16$. But recall from Definition 1.1 that a function $f(x)$ must assign to each x *precisely one value* $f(x)$. This means that $f(0)$ cannot have the two distinct values of 4 and -4. Thus the pairs $x = 0$, $y = 4$ and $x = 0$, $y = -4$ could never *both* satisfy $y = f(x)$. It follows that the equation $x^2 + y^2 = 16$ cannot be rewritten as $y = f(x)$.

So that square roots can be used in functions without giving two different values at the same point, such as $\sqrt{16} = 4$ and -4, people have agreed that the square root sign $\sqrt{}$ always means the positive square root. With this convention, writing $y = f(x)$ for $f(x) = \sqrt{16 - x^2}$ does express y as a function of x, but this equation is not quite equivalent to the original equation $x^2 + y^2 = 16$. To see this, note that the pair $x = 0$, $y = -4$ satisfies the original equation, but does not satisfy $y = \sqrt{16 - x^2}$, since -4 cannot be the *positive* square root of any number. In short, the equation $x^2 + y^2 = 16$ cannot be rewritten so that y is expressed as a function of x. The equation for the circle is $x^2 + y^2 = 16$, while $y = \sqrt{16 - x^2}$ is the equation for the top semicircle. The equation for the bottom semicircle would be $y = -\sqrt{16 - x^2}$.

Position as a Function of Time

One type of relationship you will study quite a bit, as a typical application of calculus, is that between the height above ground level of a ball thrown in the air, and the time that has elapsed since it was thrown. Typically, the letter t is used to represent time since the ball has been thrown, and the letter s (which is the first letter of *situs*, the Latin word for place or position or site) is used to represent height. Also, typically, time t is measured in seconds, and height s is measured in feet above ground level. Now the particular formula or relationship between s and t depends upon some other considerations, such as with what speed you initially throw the ball, and where you were when you threw it (on the ground, on top of a building, and so on). For now, don't worry about where the particular formula comes from (although later you will see how to get it—it comes from physics with the help of calculus).

As a specific example, if you are standing on the ground and throw a ball up into the air with an initial speed of 96 ft/s, then the height s, in feet, of the ball above the ground t seconds after you have thrown it is given by the function

$$s(t) = -16t^2 + 96t \qquad (1.3)$$

I want to make some comments about this formula. First of all, notice how much better an explicit formula is than just knowing that the ball goes up for a while and then comes back down. You can figure out exactly how high the ball is, for example, one second after you threw it by substituting $t = 1$ into the formula for $s(t)$. If you let t have the value 1, then you can compute the corresponding value of s, so that

$$s(1) = (-16 \cdot 1^2) + (96 \cdot 1) = -16 + 96 = 80$$

After 1 second, the ball is 80 feet above the ground. Secondly, notice that although the mathematical formula makes sense for all values of t, it only describes the physical situation of the thrown ball for certain values of t, namely, for t between 0 (when the ball was thrown) and 6 (when it hits the ground again, as you can check by letting $t = 6$ in the formula—when $t = 6$, $s = 0$ and the ball is back on the ground again.)

Equation 1.3 gives s as a function of t. For any time between 0 and 6 seconds, you can substitute the time value into the formula by doing what the formula tells you: square the time value, multiply the square by -16, and add to that the product of 96 multiplied by the time value. It is often very useful to think of functions as really telling you to do something to a value.

Substituting into a Function

Example 1.6 Let $f(x) = 4x^2 + 7x$. Compute $f(2), f(2+h)$, and $f(x+h)$.

> *Solution:* One more time, writing $f(x)$ means that you are thinking of f as a function of x, so that given a value of x, there is a

rule, formula, or recipe for computing the corresponding value of f. What is on the right-hand side of the equal sign is the rule. It tells you that given a value of x, you square it, multiply the square by 4, and then add to this number the product of 7 times x. This is the rule you follow, no matter what x you are given. Thus, $f(2)$ means you are given $x = 2$ to perform the recipe with, so that

$$f(2) = (4 \cdot 2^2) + (7 \cdot 2) = (4 \cdot 4) + (7 \cdot 2) = 16 + 14 = 30$$

The quantity inside the parentheses after the f is the x-value to which you apply the recipe. Another way of thinking about it is that you take whatever is inside the parentheses, and substitute that for x in the formula. Thus:

$$\begin{aligned} f(x) &= 4x^2 + 7x \\ f(?) &= 4(?)^2 + 7 \cdot ? \\ f(2) &= 4(2)^2 + 7 \cdot 2 \\ f(2 + h) &= 4(2 + h)^2 + 7(2 + h) \\ f(x + h) &= 4(x + h)^2 + 7(x + h) \end{aligned}$$

Example 1.7 With $f(x) = 4x^2 + 7x$, expand squares to simplify the expression $\dfrac{f(2 + h) - f(2)}{h}$.

Solution: Expand $f(2+h)$ first. Remember, as in Example 1.6, you apply the recipe for f to the expression inside the parentheses, which here is $2 + h$. Thus $f(2 + h) = 4(2 + h)^2 + 7(2 + h)$. Let's do the algebra in expanding $(2 + h)^2$ step by step:

$$\begin{aligned} (2 + h)^2 &= (2 + h) \cdot (2 + h) && \text{(that's what the square means)} \\ &= 2 \cdot (2 + h) + h \cdot (2 + h) && \text{(the distributive law)} \\ &= 2 \cdot 2 + 2 \cdot h + h \cdot 2 + h \cdot h && \text{(the distributive law again)} \\ &= 4 + 2 \cdot h + h \cdot 2 + h^2 \\ &= 4 + 2h + 2h + h^2 && \text{(multiplication is commutative)} \\ &= 4 + 4h + h^2 \end{aligned}$$

Note to students: If you learned about the FOIL method, you could also use that for expanding $(2 + h)^2$.

Thus:

$$\begin{aligned} f(2 + h) &= 4(2 + h)^2 + 7(2 + h) \\ &= 4(4 + 4h + h^2) + 7(2 + h) \end{aligned}$$

$$\begin{aligned} &= 16 + 16h + 4h^2 + 14 + 7h \\ &= (16 + 14) + (16h + 7h) + 4h^2 \\ &= 30 + 23h + 4h^2 \end{aligned}$$

Now clearly,

$$f(2) = 4 \cdot 2^2 + 7 \cdot 2 = 16 + 14 = 30$$

It follows that

$$f(2 + h) - f(2) = (30 + 23h + 4h^2) - 30 = 23h + 4h^2$$

Finally,

$$\frac{f(2 + h) - f(2)}{h} = \frac{23h + 4h^2}{h} = \frac{(23 + 4h) \cdot h}{h} = \frac{(23 + 4h) \cdot 1}{1} = 23 + 4h$$

1.3 GRAPHS

Equations and Graphs

In the 1600s, Rene Descartes established a connection between the subjects of algebra and geometry by introducing what we now call, in his honor, **Cartesian coordinates**. Draw intersecting horizontal and vertical lines, and mark off a unit of distance on each line, starting at the point (called the **origin**) where the lines intersect. Typically the horizontal line is called the x-axis, and the vertical line is called the y-axis. On the x-axis, the number 0 is located at the origin, and, typically, positive numbers are placed to the right, and negative numbers to the left. Likewise, on the y-axis, the number 0 is located at the origin, and, typically, positive numbers are placed above the origin, and negative numbers below. From this, you get a correspondence between all points P in the plane and all ordered pairs (x, y) of real numbers, much like the correspondence between a point on a flat map and the latitude and longitude of the point (see Figure 1.1). If a point P in the plane corresponds to an ordered pair of real numbers (x, y), the x is called the x-*coordinate* of P, and y is called the y-*coordinate* of P.

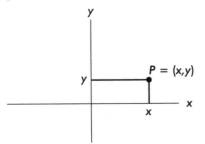

Figure 1.1. *Cartesian Coordinates*

If you have an equation such as $y = x^2$, the graph of the equation is precisely the set of all points P in the plane whose Cartesian coordinates x and y satisfy the equation. Thus the point $P = (2, 4)$ is on the graph of $y = x^2$, while the point $(2, 5)$ is not. Typically, to draw the graph of an equation, just pick a number of values for x, compute the corresponding values of y, and plot the points you have drawn. If you were either clever or lucky in picking the values of x, you can connect the dots and get a good picture of the graph. You can also, with experience, learn what the graphs of many common formulas look like. Of course, nowadays you can also use a graphing calculator.

Example 1.8 Plot the graph of the equation $x = 2$.

> *Solution:* You want to plot every point (x, y) in the plane whose x-coordinate equals 2. This means that the y-coordinate can be anything. Thus the points on the graph are the set of points of the form $(2, y)$, where y can be any number. Thus $(2, 0), (2, -1), (2, 1)$, $(2, -2), (2, 2)$, and so on are all points on the graph. Plot these points. As the y-axis is the vertical axis, the points in the plot can go arbitrarily high or low. The graph is therefore a vertical line, with all points on the line having x-coordinate equal to 2, as shown in Figure 1.2.

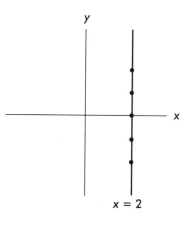

Figure 1.2. *The vertical line* $x = 2$

Often students get confused by the fact that the x-axis is horizontal, but graphs of equations such as $x = 5$ are vertical lines. Review Example 1.8 until you are sure about this. Do you see that the equation $x = 0$ has as its graph the y-axis?

Example 1.9 Plot the graph of the equation $y = -1$.

> *Solution:* Here you want to plot every point (x, y) in the plane whose y-coordinate equals -1. This means that the x-coordinate can be anything. As the x-axis is the horizontal axis, this means that points on the graph go horizontally infinitely far out to the right and to the left (a good clue that the graph is a horizontal line). Also, typical points on the graph are $(0, -1), (1, -1), (-1, -1), (2, -1), (-2, -1)$, and so on. Plot these points. All this should convince you that the graph is a horizontal line, as shown in Figure 1.3.

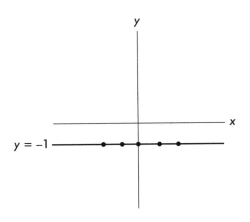

Figure 1.3. *The horizontal line* $y = -1$

Remember that the graph of an equation such as $y = 4$ is a horizontal line. The equation for the x-axis is $y = 0$.

Sometimes equations are given not with the letters x and y, but with other pairs of letters, such as t and s. That doesn't matter. You just make one of the axes correspond to one of the letters, and the other axis to the other letter. Typically, but not always, if the equation is given in the form of a function, such as $u = s(t)$, where s is a function of t, then the t-values, the values of the independent variable, would correspond to points on the horizontal axis, and the s or u values, the values of the function or the dependent variable you are computing, to points on the vertical axis.

Descartes' invention is remarkable and important because it gives you two ways of attacking a problem—through algebra or through geometry. If you have an algebraic equation you are trying to make sense of, then maybe by graphing it you can see what is going on. Conversely, if you are trying to solve a problem involving geometric objects, curves, and lines, then setting up the formulas for these objects lets you perform precise algebraic calculations.

As you study calculus, you should always be aware that for just about everything you do, there are at least two ways of looking at it—in terms of algebra and equations, and in terms of geometry and pictures in the plane. Each way of looking at a concept reinforces your understanding of it.

The Vertical Line Test

Graphically, you can easily see the difference between a general equation, and an equation that can be made into a function. Look again at Example 1.5 in Section 1.2. The example involved the equation $x^2 + y^2 = 16$, whose graph is a circle of radius 4 centered at the origin, as shown in Figure 1.4.

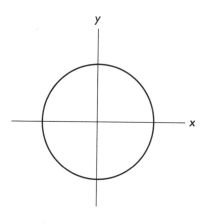

Figure 1.4. *Circle of radius 4*

If you look at the graph, you see that some vertical lines intersect the graph more than once. For example, the vertical line $x = 0$ (that is, the y-axis) intersects the graph at the point $y = 4$ and also at the point $y = -4$. This means that the graph cannot be rewritten so that y is a function of x. If y were a function of x, then you would have only one y-value for each x-value. As substituting in $x = 0$ seems to produce two values for y, namely 4 and -4, the formula $x^2 + y^2 = 16$ cannot be rewritten so that y is a function of x.

This reasoning provides the explanation for the **vertical line test**. An equation determines a function if its graph has the property that each vertical line that crosses it crosses it only once. The quantity that is plotted along the horizontal axis will represent the variable, and the quantity that is plotted along the vertical axis will represent the values of the function.

The upper semicircle in Figures 1.5 satisfies the vertical line test, and is the graph of the function $y = \sqrt{16 - x^2}$. This is because by convention the $\sqrt{}$ sign means the nonnegative square root. The lower semicircle also satisfies the vertical line test, and is the graph of the function $y = -\sqrt{16 - x^2}$.

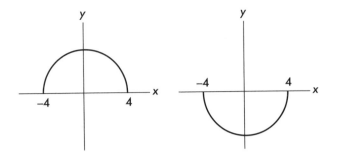

Figure 1.5. *Upper semicircle and lower semicircle*

The Domain of a Function

In the graph of $y = \sqrt{16 - x^2}$, you can see that some vertical lines do not cross the graph at all. For example, the vertical line $x = 5$ does not intersect the graph. When $x = 5$, then $16 - x^2 = -9$, and the formula for y would require computing $\sqrt{-9}$. In the system of real numbers, negative numbers have no square root, and thus the recipe for computing y when $x = 5$ makes no sense. There is no problem with this. When the equation for a function makes no sense at a particular point, the function is said to be **undefined** at that point. Functions don't have to be defined everywhere. The set of points where a function is defined (where the recipe makes sense) is called the **domain** of the function. The domain does not have to be the set of all real numbers. For most of the functions you will see in this book, the only problems with the formulas making sense will occur when either division by zero, or taking the square root of a negative number, is involved.

Example 1.10 Which of the following is the graph of a function, and which is not?

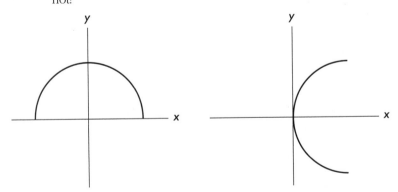

Figure 1.6

Solution: The figure on the left is the graph of a function, as it satisfies the vertical line test. Each vertical line crosses the graph in at most one point. The figure on the right is not the graph of a function. Many vertical lines to the right of the y-axis cross the graph twice.

Example 1.11 Find the domain of the function $f(x) = \dfrac{1}{x - 2}$.

Solution: You just have to worry about dividing by 0, which would occur when $x - 2 = 0$, that is, when $x = 2$. Otherwise there is no problem in carrying out the procedure specified by the equation. Thus the domain is all real numbers other than $x = 2$, or in interval notation, the domain consists of $(-\infty, 2) \cup (2, \infty)$.

1.4 SUMMARY OF MAIN POINTS

- An equation involving two quantities is a precise mathematical expression of the relationship between the two quantities.

- A function is a formula, rule, or procedure for taking a set of numbers, and assigning to each number x in the set *precisely one value*, which is denoted by $f(x)$.

- In a function, such as $f(x) = x^2 + 4$, the expression $x^2 + 4$ is the explicit statement of the procedure to be followed for taking a particular x-value and computing the corresponding value $f(x)$ of the function.

- The dependence of the values $f(x)$ upon x is expressed by saying that f is a function of x. The quantity x is thought of as the variable.

- In the notation $f(x) = x^2 + 4$, the x in parentheses can be thought of as the input to the procedure, and the $x^2 + 4$ as the rule for what you do to the input, namely, square it and add 4. Thus $f(3) = 3^2 + 4, f(-3) = (-3)^2 + 4$, and $f(x + h) = (x + h)^2 + 4$.

- Functions are closely related to special kinds of equations, in which one of the quantities stands alone on one side of the equal sign, and does not appear on the other side of the equal sign. For a function $f(x)$, the equation $y = f(x)$ is such an equation. Conversely, given such an equation, for example $y = x^2 + 4$, the right-hand side can be thought of as defining a function of x.

- The rule or procedure used in specifying a function $f(x)$ need not hold for all numbers x. The set of numbers for which it holds is called the domain of the function.

- The rule or procedure used in specifying a function $f(x)$ must specify only one value $f(x)$, for each x-value in the domain of the function.

- The $\sqrt{}$ sign, when used in equations or functions, automatically means the positive square root.

- By the use of Cartesian coordinates, equations can be graphed. This provides a connection between the subjects of algebra and of geometry.

- When graphing the equation of a function, such as $y = f(x)$ for $f(x) = x^2 + 4$, the values of the independent variable x generally correspond to points on the horizontal axis, and the corresponding values of y or $f(x)$ to points on the vertical axis.

- The vertical line test lets you see from the graph when an equation actually determines a function—when each vertical line touches the graph at most once.

- Horizontal lines have equations such as $y = 3$, and $y = -2$, while vertical lines have equations such as $x = 3$, and $x = -2$.

1.5 EXERCISES

For each of the following equations, first solve for y as a function of x, and then solve for x as a function of y. In each case, state the domain of the function.

1. $3x + 4y = 5$
2. $3xy + 2x - 1 = 0$
3. $4x^3y - 6y = 5$
4. For the function $f(x) = 2x^2 - 4x + 1$, evaluate and simplify each of the following:
   a. $f(4)$
   b. $f(-3)$
   c. $f(x + 1)$
   d. $f(2x)$
   e. $f(a)$
   f. $f(x + h)$
5. Write down an equation involving x and y for which it is easy to solve for x as a function of y, but difficult to solve for y as a function of x.
6. In freehand, draw a curve that can be thought of as the graph of a function.
7. In freehand, draw a curve that cannot be thought of as the graph of a function.

8. Sketch the graphs of each of the following functions or equations on a piece of graph paper by plotting a number of points on the graph, and then trying to reasonably connect the dots. If you have access to a graphing calculator, use it to check your graphs.

 a. $y = 2x$
 b. $y = -2x$
 c. $y = 2x + 4$
 d. $y = -2x + 4$
 e. $y = 2x - 4$
 f. $y = -2x - 4$
 g. $y = 2x^2$
 h. $y = -2x^2$
 i. $y = -2x^2 + 4$
 j. $y = \dfrac{1}{x}$
 k. $y = \sqrt{x}$
 l. $y = \sqrt{4 - x^2}$
 m. $y = 0$
 n. $x = 0$
 o. $y = 3$
 p. $x = 3$
 q. $y = -3$
 r. $x = -3$
 s. $x = 2y - 4$

2

CHANGE, AND THE IDEA OF THE DERIVATIVE

2.1 AVERAGE RATES OF CHANGE

What is calculus all about?

Calculus describes how quantities change, and uses this description of change to give us extra information about the quantities themselves. Our whole world is full of change. The height of a rocket above the surface of the earth changes as it is launched, and so does its weight as it burns up fuel; if you take medicine, the concentration of medicine in your bloodstream changes—first increasing and then decreasing; the interest rate the bank pays you for your savings deposits changes, and so do the mortgage rates for a home loan. Calculus is an essential tool in the study of all of these changes, and the examples above illustrate some of the many fields to which calculus applies—physics, medicine and biology, and economics.

More specifically, calculus gives information about the relationship between the change in the value of one quantity or item, say x, and the change in the value of another quantity or item, say y. In the examples above, you might be interested in the relationship between the change in the height of the rocket and the change in time (as time passes), or in the relationship between the change in the interest rate on savings deposits and the passage of time. You might also be interested in the relationship between the change in the interest rate on savings deposits and the change in home loan mortgage rates.

For now, call the two quantities or items you might be interested in x and y. Specific examples will be given later. How is this relationship between changes measured?

1. Find the value of the change for item x:
   (a) Find the first value of item x. Call this x_1.
   (b) Find the second value of item x. Call this x_2.
   (c) Subtract the first value x_1 from the second value x_2 to get $x_2 - x_1$. Differences or changes are computed by the operation of subtraction. This difference $x_2 - x_1$ is the value of the change in x.

2. Find the value of the change for item y:
   (a) Find the first value of item y. Call this y_1.
   (b) Find the second value of item y. Call this y_2.
   (c) Subtract the first value y_1 from the second value y_2 to get $y_2 - y_1$. This difference is the value of the change in y.

3. Find the relationship between the change in y and the change in x by dividing the change in y by the change in x to get

$$\frac{y_2 - y_1}{x_2 - x_1}$$

The above quantity, $\dfrac{y_2 - y_1}{x_2 - x_1}$, is called the **average rate of change of y with respect to x**.

Note to students: You may recognize $\dfrac{y_2 - y_1}{x_2 - x_1}$ as the definition of the slope of a line. The slope of a line is just *one* example of an average rate of change of y with respect to x.

The word **rate** is related to the word *ratio*, which means a fraction, and thus rates generally involve fractions and division of one quantity by another. The phrase **rate of change** means that you are describing the rate, or ratio, of a change in one quantity with the change in another quantity, so that you are dividing the change in one quantity by the change in another quantity. The phrase **of y with respect to x** *always* means that the change in y is the numerator, or the top part, of the fraction, and that the change in x is the denominator, or bottom part, of the fraction, so that you are dividing the change in y by the change in x. The rate of change of x with respect to y would be the change in x divided by the change in y, that is, $\dfrac{x_2 - x_1}{y_2 - y_1}$. The word **average** serves to distinguish this type of rate of change from another, which you will learn about in Section 2.2 of this chapter.

If y and x represent particular named quantities, often the phrase *rate of change of y with respect to x* is awkward. A substitute is *rate of change in y per x,* or something similar. The word *per* is a substitute for *with respect to*. For example, velocity means the rate of change of position with respect to time. But the common way of talking about the velocity of a car is in terms of *miles per hour*, and not as *the rate of change in miles with respect to hours*.

You should notice, from Step 1(c) and Step 2(c) on page 17 that the change in an item is always computed as the second value of the item minus the first value. Thus the change has a sign, which tells you whether the item is getting larger or smaller as it changes. For example, if you first see that a ball is 5 feet above the ground, and then see that it is 8 feet above the ground, then its change in height is $8 - 5 = 3$ feet. The positive change tells you that the ball has increased its height above the ground, so that it is going up. If you first see that a ball is 8 feet above the ground, and then a little bit later see that it is 5 feet above the ground, then its change in height is $5 - 8 = -3$ feet. The negative change tells you that the height of the ball above the ground has decreased, so that the ball is falling.

Example 2.1 A man weighed 160 pounds at his last physical exam exactly 3 years ago, and at his exam today he weighed 185 pounds. (a) What is his average rate of change, in pounds per year, between the previous physical exam and the current one? (b) What is his average rate of change in pounds per month over the same period?

Solution: (a) The phrase *average rate of change in pounds per year* means the ratio of the change in pounds divided by the change in years. The change in pounds is computed by subtracting the first weight value from the second, to get $185 - 160 = 25$. Although the specific years of his exams are not given, the change in years is clearly 3. Thus the average rate of change in pounds per year is the ratio $\dfrac{25}{3} = 8\dfrac{1}{3}$ lbs/yr. (b) The change in months is clearly $12 \cdot 3 = 36$. Thus the average rate of change in pounds per month is the ratio of the change in pounds divided by the change in months, which is $\dfrac{25}{36} = 0.694$ lb/mo.

Example 2.2   A woman on a dieting program lost weight at the average rate of 2 pounds per month. If she initially weighed 132 pounds and now weighs 110 pounds, for how many months has she been on the program?

Solution: You can probably easily solve this problem in your head, but also read how to do it by using the rule:

$$\text{average rate of change in pounds per month} = \frac{\text{change in pounds}}{\text{change in months}}$$

Working an easier problem *systematically* helps you see how to use the same system to work harder problems. In the above rule, you know two out of the three quantities, so you can use algebra to find the third quantity. The change in pounds is $110 - 132 = -22$. Remember, you always compute change by subtracting the earlier value from the later value. The negative sign here indicates that the woman lost weight. Also, since the woman *lost* 2 pounds per month, on average, her average rate of change in pounds per month is -2. Thus,

$$-2 = \frac{-22}{\text{change in months}}$$

so that

$$\text{change in months} = \frac{-22}{-2} = 11$$

The woman has been on the dieting program for 11 months.

Average Velocity

One relationship that is often measured is that between the change in position s (of some object) and the change in time t. The average rate of change of position with respect to time comes up so often that there is a special name for it, **average velocity**, v. Whenever you see the word *velocity*, think *rate of change of position with respect to time*.

To compute the average velocity of a moving ball, suppose the ball has position s_1 at time t_1 and position s_2 at time t_2. Then its average velocity (between times t_1 and t_2) is given by:

$$\frac{s_2 - s_1}{t_2 - t_1}$$

To review, this is because average velocity means average rate of change of position with respect to time, which means the change (or difference—computed by subtracting the first value from the second) in position divided by the change (or difference, likewise computed by subtracting the first value from the second) in time.

Example 2.3 A ball is thrown into the air. It is observed that 1 second after being thrown it is 80 feet high, and 3 seconds after being thrown, it is 144 feet high. Find the average velocity of the ball during the time period between 1 and 3 seconds after it was thrown.

Solution: As average velocity means average rate of change of position with respect to time, you have to divide the change in position by the change in time. You must first determine the two time values t_1 and t_2, and the two corresponding position values s_1 and s_2. From the wording of the example, you can see that the first time is $t_1 = 1$ and the corresponding first position is $s_1 = 80$, while the second time is $t_2 = 3$ and the corresponding second position is $s_2 = 144$. Thus,

$$\frac{s_2 - s_1}{t_2 - t_1} = \frac{144 - 80}{3 - 1} = \frac{64}{2} = 32 \text{ feet per second}$$

Example 2.4 For the same ball as in Example 2.3, it is observed that 4 seconds after being thrown, it is 128 feet high. Find the average velocity of the ball during the time period between 3 and 4 seconds after it was thrown.

Solution: Now the first time is $t_1 = 3$, with corresponding position $s_1 = 144$, from Example 2.3, and the second time is $t_2 = 4$, with corresponding position $s_2 = 128$. Thus the average velocity equals

$$\frac{s_2 - s_1}{t_2 - t_1} = \frac{128 - 144}{4 - 3} = \frac{-16}{1} = -16 \text{ feet per second}$$

Here the negative average velocity means that the ball has decreased in height, so you know that it has fallen.

Example 2.5   A car driving due east away from Atlanta is 20 miles from the city limits at 1 P.M. and 130 miles from the city limits at 3 P.M. What is the car's average velocity between 1 P.M. and 3 P.M.?

Solution: Position can be measured by measuring the distance to any fixed point. The fixed point in this example is the city limits of Atlanta. Thus at time $t_1 = 1$, the first position of the car is $s_1 = 20$, and at time $t_2 = 3$, the second position of the car is $s_2 = 130$. The average velocity of the car is thus

$$\frac{s_2 - s_1}{t_2 - t_1} = \frac{130 - 20}{3 - 1} = \frac{110}{2} = 55 \text{ miles per hour, away from Atlanta}$$

Example 2.6   A ball thrown up into the air has an average velocity, between 3 seconds and 5 seconds after it was thrown, of -14 feet per second. If 5 seconds after it was thrown the ball was 20 feet above the ground, how high was it 3 seconds after it was thrown?

Solution: You still use the equation

$$\text{Average velocity} = \frac{\text{change in position}}{\text{change in time}} = \frac{s_2 - s_1}{t_2 - t_1}$$

Unlike the examples above, in this example you are not given direct information for all of the four quantities t_1, t_2, s_1, and s_2. Thus you cannot directly compute the ratio on the right-hand side of the equation. That does not matter. Of the five quantities t_1, t_2, s_1, s_2 and *average velocity*, you are given four of them. You can use the equation and a little algebra to get the fifth. What is given is $t_1 = 3, t_2 = 5, s_2 = 20$ and average velocity $= -14$. You want to solve for s_1. Thus,

$$-14 = \frac{20 - s_1}{5 - 3} = \frac{20 - s_1}{2}$$

so that

$$\begin{aligned} 2 \cdot -14 &= 20 - s_1 \\ -28 &= 20 - s_1 \end{aligned}$$

and

$$s_1 = 20 + 28 = 48 \text{ feet}$$

Average Rate of Change of a Function

Generally, in computing average rates of change of a quantity y with respect to a quantity x, there is a function that shows how the values of x and of y are related. For example, suppose $y = f(x) = 16x^2$. (Recall the discussion starting on page 3 for an explanation of the notation $y = f(x)$.) For each value of x, there is a corresponding value of y or $f(x)$, found by substituting the value for x into the formula $16x^2$. If you had two values of x, say x_1 and x_2, the standard notations for the two corresponding values of y or $f(x)$ would be either y_1 and y_2, or $f(x_1)$ and $f(x_2)$. It would make perfectly good sense to denote the two corresponding values also by f_1 and f_2, but this is generally not done.

Example 2.7 Let $y = 16x^2$. Compute the average rate of change of y with respect to x, as x changes from 2 to 3.

Solution: The first value of x mentioned is $x_1 = 2$, and the second value of x mentioned is $x_2 = 3$. The corresponding y values are

$$y_1 \;=\; f(x_1) = f(2) = 16 \cdot 2^2 = 64$$

and

$$y_2 \;=\; f(x_2) = f(3) = 16 \cdot 3^2 = 144$$

The average rate of change of y with respect to x is then

$$\frac{\text{the change in } y}{\text{the change in } x} = \frac{y_2 - y_1}{x_2 - x_1} = \frac{144 - 64}{3 - 2} = \frac{80}{1} = 80$$

Example 2.8 A ball thrown straight up into the air has a height above the ground of $s(t) = -16t^2 + 96t$ feet, t seconds after it was thrown. Find the average velocity of the ball during the time period between 1 and 3 seconds after it was thrown. (Some common faster ways of stating the same problem are "Find the average velocity of the ball between times 1 and 3"; and "Find the average velocity of the ball between $t = 1$ and $t = 3$.")

Solution: At the first time, $t_1 = 1$, the height of the ball is

$$s_1 = s(t_1) = s(1) = -16 \cdot 1^2 + 96 \cdot 1 = -16 + 96 = 80 \text{ feet}$$

At the second time, $t_2 = 3$, the height of the ball is

$$s_2 = s(t_2) = s(3) = -16 \cdot 3^2 + 96 \cdot 3 = -144 + 288 = 144 \text{ feet}$$

Thus the average velocity, the average rate of change of position with respect to time, is

$$\frac{s_2 - s_1}{t_2 - t_1} = \frac{144 - 80}{3 - 1} = \frac{64}{2} = 32 \text{ feet per second}$$

Compare Example 2.8 with Example 2.3. In Example 2.3, the two times and the two positions were given explicitly. In Example 2.8, only the two times had to be given explicitly, since given a time value for t, you can compute the height of the ball yourself using the formula for $s(t)$. Otherwise the two examples are identical.

2.2 INSTANTANEOUS RATES OF CHANGE

Think for a few minutes about what the title of this section could possibly mean.

Are you a little puzzled? I hope so! Doesn't the phrase "instantaneous rate of change" seem like an oxymoron, a contradiction in terms (like the phrase "thunderous silence")? The part of the phrase "rate of change" implies that two quantities are changing, and this change must require some time to occur. On the other hand, the word "instantaneous" means that no time is passing. How can a quantity change from one value to another instantly?

Instantaneous Velocity—The Concept

Here's an example that should be familiar to everyone: You are driving along a straight road in your car, glance down at your speedometer, and see that at the very instant you glanced at it, your speed, or velocity, was 40 miles per hour. You don't necessarily assume that a few seconds later your velocity will still be exactly 40 miles per hour—you may be speeding up or slowing down. This situation is so familiar that its meaning is intuitively understood—your velocity at a certain instant is 40 miles per hour. But how does this relate to your understanding of velocity as measuring change in position over change in time? You don't necessarily expect that for the rest of your trip, your change in position (miles traveled), divided by change in time (how long the trip takes) will equal 40. From experience, you know that driving for another half hour will not automatically mean you will have driven 20 more miles. It would if your velocity remained the same, at 40 miles per hour. But over the next half hour, you may speed up or slow down many times, and reading your speedometer now really gives you very little information as to how far you will travel in the next half hour.

Perhaps it seems more reasonable to expect that reading your speedometer now will at least give you good information about how far you will travel in the next minute ($\frac{1}{60}$ of an hour). If you could use a stopwatch and your odometer to compute the distance you travel in the next minute, and express your average velocity,

$$\frac{\text{change in position}}{\text{change in time}} = \frac{\text{distance traveled}}{1 \text{ minute}}$$

in miles per hour (by changing units), you might expect to get an answer close to the 40 miles per hour your speedometer showed when you glanced at it. Would you really expect this? Well, it depends. If nothing much changes in road or traffic conditions, and you don't change your driving style much, maybe. But what if, a few seconds after you glanced at your speedometer, you came to a red light, and stopped for the rest of the minute?

Well, maybe even a minute is a long time when driving. Conditions can change dramatically in a minute. Suppose I put a physicist, with highly accurate measuring devices for time and position, along the side of the road, glance down at my speedometer at the instant I pass her, and ask her to measure the distance I travel in the next second, divide the distance by the change in time (1 second), and change units so the rate of change of position with respect to time is given in miles per hour. Barring some disaster (like crashing into a car stopped ahead of me, half a second after I pass her), I should get an answer that's pretty close to what my speedometer read at the instant I glanced at it.

Even if I crash into the car stopped directly ahead of me, I can give meaning to the phrase "my velocity at this instant is 40 miles per hour," in terms of change in position over change in time. I just can't use a time interval as long as 1 second. Why not use $\frac{1}{1000}$ of a second? Remember, the physicist has highly accurate measuring devices. The distance I travel over the next $\frac{1}{1000}$ of a second, divided by the change in time ($\frac{1}{1000}$ of a second), converted to miles per hour, gives my average velocity, in miles per hour, during the time interval of $\frac{1}{1000}$ of a second from the time I pass the physicist and glance down at my speedometer. Don't you think it would be very surprising if this number were not very, very close to the 40 miles per hour I read on my speedometer?

The point of the whole discussion above is that, because of your experience with traveling and looking at speedometers, both the concept of *average velocity* and the concept of *velocity at an instant* (instantaneous velocity, or instantaneous rate of change of position with respect to time) have intuitive meaning to you. The connection between the two concepts is that if you compute the average velocity over smaller and smaller time periods (from the instant you look at the speedometer), you should get numbers that are closer and closer to the speedometer reading at the instant you looked at it, your instantaneous velocity.

Now the oxymoron can be explained! What does the title of this section mean?

DEFINITION 2.1. *Instantaneous rate of change is defined as "what happens to average rates of change, as you compute the average over smaller and smaller intervals."*

Instantaneous Velocity—A Numerical Approach

Suppose a ball is thrown straight up into the air, and the following table gives information about the height s of the ball above ground, in feet, t seconds after it was thrown:

t	s	t	s
1	80	3	144
1.9	124.64	2.1	131.04
1.99	127.6784	2.01	128.3184
1.999	127.967984	2.001	128.031984
2	128	2	128

Suppose you want to know the velocity of the ball at the instant $t = 2$. Can you figure this out from the table? Remember, velocity means rate of change of position with respect to time, so velocity at the instant $t = 2$ is just the instantaneous rate of change of position with respect to time, at $t = 2$. By Definition 2.1, you should be able to figure this out by computing average rates of change of position with respect to time, for smaller and smaller intervals of time. Since one second can make a big difference in the velocity or position of a ball thrown up into the air, you might not expect the average velocity over a time interval of one second to give much information about instantaneous velocity, but you should expect average velocity over a time interval of $\frac{1}{1000}$ of a second to give pretty good information.

From the table, you can compute the following:
average velocity between times 2 and 3:

$$\frac{144 - 128}{3 - 2} = \frac{16}{1} = 16 \text{ feet per second}$$

average velocity between times 2 and 2.1:

$$\frac{131.04 - 128}{2.1 - 2} = \frac{3.04}{0.1} = 30.4 \text{ feet per second}$$

average velocity between times 2 and 2.01:

$$\frac{128.3184 - 128}{2.01 - 2} = \frac{0.3184}{0.01} = 31.84 \text{ feet per second}$$

average velocity between times 2 and 2.001:

$$\frac{128.031984 - 128}{2.001 - 2} = \frac{0.031984}{0.001} = 31.984 \text{ feet per second}$$

Look at these average velocities over smaller and smaller time intervals past time $t = 2$:

$$16 \quad 30.4 \quad 31.84 \quad 31.984$$

It certainly looks like these numbers are getting closer and closer to 32, and you might reasonably guess that the instantaneous velocity of the ball at time $t = 2$ is 32 feet per second.

Instantaneous Velocity—An Algebraic Approach

In Example 2.8, you were told that a certain ball thrown up into the air has height $s(t) = -16t^2 + 96t$ feet, t seconds after it was thrown. If you use a calculator and plug in the values

$$t = 1, \ 1.9, \ 1.99, \ 1.999, \ 2, \ 2.001, \ 2.01, \ 2.1, \text{ and } 3$$

into the above formula for $s(t)$, you will see that you get precisely the s values in the table on page 25. Once you know a formula that gives the height s at all times t, there is another way to get the instantaneous velocity at $t = 2$. You don't have to pick a bunch of time values getting closer and closer to 2, such as $t = 3$, $t = 2.1$, $t = 2.01$, and $t = 2.001$, compute the average velocities between $t = 2$ and each of these other times, and then try to make a good guess about what is happening to these average velocities. You can use mathematical abstraction, and letters standing for numbers (that is, algebra) to compute all of the average velocities over small time periods, all at once.

Think of h as standing for a small positive number, or a small interval of time, so that $2 + h$ is a time a little bit after time 2. To compute the average velocity of the ball between times $t_1 = 2$ and $t_2 = 2 + h$, you need to compute first the corresponding positions s_1 and s_2:

$$s_1 \ = \ s(t_1) = s(2) = -16 \cdot 2^2 + 96 \cdot 2 = 128$$

$$
\begin{aligned}
s_2 \ = \ & s(t_2) = s(2 + h) = -16 \cdot (2 + h)^2 + 96 \cdot (2 + h) \\
= \ & -16 \cdot (4 + 4h + h^2) + 96 \cdot (2 + h) \\
= \ & (-64 + 192) + (-64h + 96h) - 16h^2 \\
= \ & 128 + 32h - 16h^2
\end{aligned}
$$

The change in position is thus

$$s_2 - s_1 = (128 + 32h - 16h^2) - 128 = 32h - 16h^2$$

while the change in time is

$$t_2 - t_1 = (2 + h) - 2 = h$$

So, the average velocity is

$$\frac{s_2 - s_1}{t_2 - t_1} = \frac{32h - 16h^2}{h}$$

$$= \frac{32h}{h} - \frac{16h^2}{h}$$

$$= 32 - 16h$$

Thus the *average* velocity between times $t = 2$ and $t = 2 + h$ is given by the formula

$$32 - 16h \tag{2.1}$$

As a check that the formula is correct, suppose for example that

$$t_2 = 2 + h = 2.001, \text{ so that } h = 0.001$$

Then the average velocity, by Equation 2.1, is

$$32 - (16 \cdot 0.001) = 32 - 0.016 = 31.984$$

This is the same value obtained from the calculations using the table on page 25.

Now comes the fascinating part. Remember, the goal is to compute the instantaneous velocity at $t = 2$, that is, to compute what happens to the average velocities over smaller and smaller time periods, starting at $t = 2$. The length of the time period from $t_1 = 2$ to $t_2 = 2 + h$ is just h, so the time period is getting smaller and smaller precisely as h gets closer and closer to 0. As the average velocity from $t_1 = 2$ to $t_2 = 2 + h$ is $(32 - 16h)$, from Equation 2.1, you just have to compute what happens to the values of $(32 - 16h)$ as h gets closer and closer to 0. Hopefully, it seems reasonable to you that as h gets closer and closer to 0, then the product $-16 \cdot h$ gets closer and closer to $-16 \cdot 0 = 0$, so that the difference $32 - 16h$ gets closer and closer to $32 - 16 \cdot 0 = 32 - 0 = 32$. Thus the instantaneous velocity at $t = 2$ equals 32 feet per second, which is exactly what was obtained on page 26.

Note to students: The algebraic technique for easily seeing what happens as h gets closer and closer to 0 was expanding the square in computing s_2. This allows you to cancel some terms in computing $s_2 - s_1$. After cancellation, all the remaining terms had at least one h in them, so that you could really divide out by $t_2 - t_1 = h$ in computing average velocity. Be on the lookout for this same technique in subsequent examples.

There is a special name and special symbolism for describing the technique of looking at an expression such as $32 - 16h$, and seeing what happens to its values as one of the variables, or letters, in the expression, such as h, gets closer and closer to a particular value, such as 0. It is called **finding a limit**. Symbolically, it is written

$$\lim_{h \to 0} (32 - 16h) = 32$$

The "lim" of course stands for the word "limit." Beneath the "lim," the symbol $h \to 0$ means that the quantity h is changing, and that it is getting closer and closer to 0. The whole left-hand side stands for the procedure of figuring out what happens to the values of $32 - 16h$ as h gets closer and closer to 0, and for the particular number that these values are getting closer and closer to. The equal sign, and the 32 on the right-hand side, mean that 32 is that particular number.

We will study limits more systematically in the next chapter, but, hopefully, for now you understand the notation well enough to follow along.

Example 2.9 For the ball with height $s(t) = -16t^2 + 96t$, find the average velocity of the ball between times $t = 1$ and $t = 1 + h$.

Solution: At the first time, $t_1 = 1$, the height of the ball is

$$s_1 = s(t_1) = s(1) = -16 \cdot 1^2 + 96 \cdot 1 = -16 + 96 = 80$$

At the second time, $t_2 = 1 + h$, the height of the ball is

$$
\begin{aligned}
s_2 = s(t_2) = s(1 + h) &= -16(1 + h)^2 + 96(1 + h) \\
&= -16(1 + 2h + h^2) + 96(1 + h) \\
&= -16 - 32h - 16h^2 + 96 + 96h \\
&= (-16 + 96) + (-32h + 96h) - 16h^2 \\
&= 80 + 64h - 16h^2
\end{aligned}
$$

The change in position is thus

$$s_2 - s_1 = (80 + 64h - 16h^2) - 80 = 64h - 16h^2$$

while the change in time is

$$t_2 - t_1 = (1 + h) - 1 = h$$

so, the average velocity is

$$
\begin{aligned}
\frac{s_2 - s_1}{t_2 - t_1} &= \frac{64h - 16h^2}{h} \\
&= \frac{64h}{h} - \frac{16h^2}{h} \\
&= 64 - 16h
\end{aligned}
$$

Example 2.10 Use Example 2.9 to find the instantaneous velocity of the ball at time $t = 1$.

> **Solution:** Example 2.9 says that the average velocity of the ball from time $t = 1$ to time $t = 1 + h$ is given by $64 - 16h$. To compute what happens to these values as h gets closer and closer to 0, observe (you can experiment with your calculator if you like) that it seems reasonable to suppose that $16h$ gets closer and closer to $16 \cdot 0 = 0$, so that $64 - 16h$ gets closer and closer to $64 - 0 = 64$. Thus 64 feet per second is the instantaneous velocity of the ball at time $t = 1$. In other words, the instantaneous velocity of the ball at time $t = 1$ is given by
>
> $$\lim_{h \to 0} (64 - 16h) = 64 - 16 \cdot 0 = 64$$

Instantaneous Velocity—An Even More Algebraic Approach

The above approach to computing instantaneous velocity is more abstract than the numerical approach based upon the table on page 25, using as it does algebra rather than just arithmetic. The approach you will see now is a little more abstract and algebraic, but it yields a tremendous advantage—with the formula $s(t) = -16t^2 + 96t$ for position as a function of time, it yields one formula for the instantaneous velocity as a function of time t. You may have noticed that the work done in Examples 2.9 and 2.10 to compute the instantaneous velocity at time $t = 1$ was very similar to the prior discussion on computing the instantaneous velocity at time $t = 2$. To compute instantaneous velocity at a general time t, all you have to do is replace the 1 or 2 value of t with the letter t, standing for a general time value.

Note to students: If the algebra below is troublesome for you, there is a brief algebra review in Chapter 4, "Computing Some Derivatives."

Let the first time value be $t_1 = t$, and the second time value $t_2 = t + h$. The corresponding first position is

$$s_1 = s(t_1) = s(t) = -16t^2 + 96t$$

The corresponding second position of the ball is

$$
\begin{aligned}
s_2 = s(t_2) = s(t + h) &= -16(t + h)^2 + 96(t + h) \\
&= -16(t^2 + 2th + h^2) + 96(t + h) \\
&= -16t^2 - 32th - 16h^2 + 96t + 96h
\end{aligned}
$$

Thus the change in position of the ball is

$$s_2 - s_1 = (-16t^2 - 32th - 16h^2 + 96t + 96h) - (-16t^2 + 96t)$$

The terms $-16t^2$ and $96t$ cancel to give

$$s_2 - s_1 = -32th - 16h^2 + 96h$$

as the change in position, while the change in time is

$$t_2 - t_1 = (t + h) - t = h$$

Thus, the average velocity between times $t_1 = t$ and $t_2 = t + h$ is:

$$\frac{s_2 - s_1}{t_2 - t_1} = \frac{-32th - 16h^2 + 96h}{h}$$
$$= \frac{-32th}{h} - \frac{16h^2}{h} + \frac{96h}{h}$$
$$= -32t - 16h + 96$$

To see what happens to this average velocity over smaller and smaller time intervals, that is, as h gets closer and closer to 0, observe that it certainly seems reasonable to expect that $-16h$ should get closer and closer to $-16 \cdot 0 = 0$, and thus $-32t - 16h + 96$ should get closer and closer to $-32t - 0 + 96 = -32t + 96$. You now have the instantaneous velocity at time t expressed as a function of t. Using v for instantaneous velocity,

$$v(t) = -32t + 96 \tag{2.2}$$

Note that at time $t = 2$,

$$v(2) = -32 \cdot 2 + 96 = -64 + 96 = 32 \text{ feet per second}$$

At time $t = 1$,

$$v(1) = -32 \cdot 1 + 96 = -32 + 96 = 64 \text{ feet per second}$$

These are the same answers obtained earlier.

It took some work to get Equation 2.2, but look at what you've got. If you want to compute instantaneous velocity at some other time, say $t = 4$, you don't have to experimentally determine a table of positions and times, as on page 25, and do a lot of numerical calculations. You don't even have to do the algebra like that done in Examples 2.9 and 2.10. All you have to do is substitute 4 for the t-value in Equation 2.2 to get

$$v(4) = -32 \cdot 4 + 96 = -128 + 96 = -32 \text{ feet per second}$$

Note to students: In everyday language, the terms "speed" and "velocity" are often used interchangeably, but there is an important technical distinction between the two. The term "velocity" means the rate of change of position with respect to time, and so can be positive, if the position is *increasing*, or getting larger, with respect to time, or negative, if the position is *decreasing*, or getting smaller, with respect to time. The term "speed" is by definition *always* positive, and is simply defined as the absolute value of the velocity. In the ball problems, we've used the common convention that up is the positive direction and down is the negative direction. Thus, the fact that $v(4) = -32$ tells you not only how fast the ball is traveling at the instant $t = 4$, but also that its height function s is getting smaller at that instant. That is, at time $t = 4$ the ball is falling. While the velocity of the ball is -32 feet per second, the speed of the ball is 32 feet per second. In everyday language, you would say that at time $t = 4$ the ball is *falling* at the speed of 32 feet per second.

Example 2.11 Suppose that a bowling ball is $s(t) = 45t + 2$ feet beyond the foul line t seconds after it is let go. Find a formula for the instantaneous velocity $v(t)$ of the ball at time t.

Solution: You have to compute the average velocity of the ball between times $t_1 = t$ and $t_2 = t + h$, and then determine what happens to these average velocities as h gets closer and closer to 0. At time $t_1 = t$, the position of the ball is

$$s(t_1) = s(t) = 45t + 2$$

At time $t_2 = t + h$, the position of the ball is

$$s(t_2) = s(t + h) = 45(t + h) + 2 = 45t + 45h + 2$$

So, the change in position of the ball is

$$s_2 - s_1 = (45t + 45h + 2) - (45t + 2) = 45h$$

The change in time is

$$t_2 - t_1 = (t + h) - t = h$$

The average velocity between times t and $t + h$ is thus

$$\frac{s_2 - s_1}{t_2 - t_1} = \frac{45h}{h} = 45 \text{ feet per second}$$

As the formula for average velocity between times t and $t + h$ has no h in it, there is nothing else to determine as h gets closer and closer to 0. The instantaneous velocity is also 45 feet per second. In this problem, the speed of the ball is constant, and thus all average velocities and instantaneous velocities are the same.

Example 2.12 (a) Find the instantaneous velocity $v(t)$ of a ball thrown up into the air, if its height above ground at time t is given by the formula $s(t) = -16t^2 + 144t$. (b) Find the instantaneous velocity of the ball at time $t = 2$.

Solution: (a) At time $t_1 = t$, the position of the ball is

$$s(t_1) = s(t) = -16t^2 + 144t$$

At time $t_2 = t + h$, the height of the ball is

$$
\begin{aligned}
s_2 = s(t_2) = s(t + h) &= -16(t + h)^2 + 144(t + h) \\
&= -16(t^2 + 2th + h^2) + 144(t + h) \\
&= -16t^2 - 32th - 16h^2 + 144t + 144h
\end{aligned}
$$

Thus the change in position of the ball is given by

$$s_2 - s_1 = (-16t^2 - 32th - 16h^2 + 144t + 144h) - (-16t^2 + 144t)$$

The terms $-16t^2$ and $144t$ cancel to give the change in position as

$$s_2 - s_1 = -32th - 16h^2 + 144h$$

The change in time is

$$t_2 - t_1 = (t + h) - t = h$$

Thus, the average velocity between times $t_1 = t$ and $t_2 = t + h$ is

$$
\begin{aligned}
\frac{s_2 - s_1}{t_2 - t_1} &= \frac{-32th - 16h^2 + 144h}{h} \\
&= \frac{-32th}{h} - \frac{16h^2}{h} + \frac{144h}{h} \\
&= -32t - 16h + 144
\end{aligned}
$$

To see what happens to this average velocity over smaller and smaller time intervals, that is, as h gets closer and closer to 0, observe that $-16h$ will get closer and closer to 0, so that $-32t - 16h + 144$ will get closer and closer to $-32t - 0 + 144 = -32t + 144$. Thus the instantaneous velocity $v(t) = -32t + 144$.

(b) Now that you already have a formula for the instantaneous velocity of the ball at any time, the instantaneous velocity at time $t = 2$ is simply $v(2) = -32 \cdot 2 + 144 = 80$ feet per second.

2.3 THE DERIVATIVE

Let $f(x)$ be a function of x. If you want to keep a particular example in mind, try $f(x) = -16x^2 + 96x$. As the values of x change, so in general do the values $f(x)$. Thus you can think about *average* rates of change of f with respect to x, and *instantaneous* rates of change of f with respect to x.

Recall from Section 2.1 that an average rate of change is a ratio or fraction, given by a change in one quantity divided by a change in another quantity. The average rate of change of f with respect to x means the change in f divided by the change in x. Two different values of x are given, and the change in x is the difference in the two values, determined by subtracting the first value from the second. As f is a function of x, each value of x determines a corresponding value of f, and the change in f is the difference in the two corresponding values of f. Finally, recall from Section 2.2 that an instantaneous rate of change of f with respect to x is simply a limit of average rates of change over smaller and smaller intervals.

To get an equation for the *derivative of f at x*, think of one value of x as fixed at the value x_1. The second, different value of x will be $x_2 = x_1 + h$. This notation emphasizes how the second value has changed from the first value. Thus the change in x is

$$x_2 - x_1 = (x_1 + h) - x_1 = h$$

The two corresponding values of f are

$$f(x_1)$$

and

$$f(x_2) = f(x_1 + h)$$

so that the change in f is

$$f(x_2) - f(x_1) = f(x_1 + h) - f(x_1)$$

Thus, the average rate of change of f with respect to x, as x changes from x_1 to $x_2 = x_1 + h$, is the ratio

$$\frac{f(x_2) - f(x_1)}{x_2 - x_1} = \frac{f(x_1 + h) - f(x_1)}{(x_1 + h) - x_1} = \frac{f(x_1 + h) - f(x_1)}{h}$$

The instantaneous rate of change of f with respect to x, at x_1, is the limit of the above average rates of change, as the change h in x gets closer and closer to 0. Thus, using the limit notation, the instantaneous rate of change of f with respect to x, at x_1, is

$$\lim_{h \to 0} \frac{f(x_1 + h) - f(x_1)}{h}$$

The **derivative of f with respect to x, at $x = x_1$**, is nothing more than the instantaneous rate of change of f with respect to x, at $x = x_1$, and is thus given by the above formula. Now just drop the subscript 1 from the x in the above formula, and you'll get the instantaneous rate of change of f with respect to x, at a general point x. This is called the **derivative of f at x**.

DEFINITION 2.2. *The **derivative of f at x** equals*

$$\lim_{h \to 0} \frac{f(x + h) - f(x)}{h}$$

Derivatives, or instantaneous rates of change, and their applications are the subject of that part of calculus called differential calculus. From now on, the phrase "rate of change" of f with respect to x will automatically mean "instantaneous rate of change" of f with respect to x, or, in mathematical terms, the derivative of f at x, given by Definition 2.2. If you change the letter x to t, and the letter f to s, and think of $s(t)$ as giving the position of an object at time t, then the (instantaneous) velocity $v(t)$ is simply the derivative of s at t.

Some Notation

The mathematical symbolism for "derivative of f at x" is $f'(x)$, which is read as "f prime of x." Thus

$$f'(x) = \lim_{h \to 0} \frac{f(x + h) - f(x)}{h} \tag{2.3}$$

In Equation 2.2, with the position function

$$s(t) = -16t^2 + 96t$$

you obtained

$$v(t) = s'(t) = -32t + 96$$

There are other commonly used notations for the derivative. Often, a change in a quantity q is expressed by the symbol Δq. (The symbol Δ is the capital Greek letter delta.) You should not think of this as a Δ times a q, but rather as one quantity, the change in q. When this notation is used with a general function $f(x)$, it is common to think in terms of $y = f(x)$ and to denote the change in the values of the function by Δy rather than Δf. For functions that refer to particular quantities, such as $s(t)$ referring to position as a function of time, it is common to use the delta notation for a change in the value of the function. So, you will commonly see a change in position denoted by Δs.

Let $y = f(x)$. The average rate of change of y with respect to x can be symbolized as

$$\frac{\Delta y}{\Delta x} \tag{2.4}$$

This is just the change in y divided by the change in x. The change in x, denoted before as h, is now denoted as Δx. With this notation, you have

$$f'(x) = \lim_{\Delta x \to 0} \frac{\Delta y}{\Delta x}$$

The Δ notation for change motivates a second common notation for the derivative, namely, $\frac{dy}{dx}$. You can think of the "d" as what becomes of the Δ as Δx gets closer and closer to 0. Thus,

$$\frac{dy}{dx} = \lim_{\Delta x \to 0} \frac{\Delta y}{\Delta x}$$

For the function $s(t) = -16t^2 + 96t$,

$$v(t) = \frac{ds}{dt} = -32t + 96$$

For $y = f(x)$, other common notations for the derivative are $\frac{df}{dx}$ or y'.

Example 2.13 Find the derivative $f'(x)$ of $f(x) = -16x^2 + 144x$.

Solution: Use Equation 2.3. You must first expand the expression for $f(x + h)$ to allow cancellations when subtracting $f(x)$.

$$
\begin{aligned}
f(x + h) &= -16(x + h)^2 + 144(x + h) \\
&= -16(x^2 + 2xh + h^2) + 144(x + h) \\
&= -16x^2 - 32xh - 16h^2 + 144x + 144h
\end{aligned}
$$

Thus,

$$
\begin{aligned}
f(x + h) - f(x) &= (-16x^2 - 32xh - 16h^2 + 144x + 144h) - (-16x^2 + 144x) \\
&= -32xh - 16h^2 + 144h
\end{aligned}
$$

Finally,

$$
\begin{aligned}
f'(x) &= \lim_{h \to 0} \frac{f(x + h) - f(x)}{h} \\
&= \lim_{h \to 0} \frac{-32xh - 16h^2 + 144h}{h} \\
&= \lim_{h \to 0} (-32x - 16h + 144) \\
&= -32x - 16 \cdot 0 + 144 \\
&= -32x + 144
\end{aligned}
$$

Applications of the Derivative—A Preview

What's the big deal about derivatives? Surprisingly enough, even if you have a formula for a quantity, knowing about how the quantity is changing can give you extra information that is not so obvious from the formula. Let's go back to the example of the ball being thrown up into the air, with height $s(t) = -16t^2 + 96t$ feet after t seconds. How high does the ball go? It is not so easy to answer this directly from the formula for height $s(t)$. However, the formula for velocity $v(t) = -32t + 96$ will be helpful. Recall that when the ball has positive velocity, it is going up, because its height is increasing in time, and when the ball has negative velocity, it is falling, because its height is decreasing in time. At the instant it reaches its high point, it is changing from rising to falling, and its velocity is changing from being positive (but close to 0) to being negative (but close to 0). At that *instant* it is sort of just hanging there, and at that *instant* its velocity is exactly 0. Upon setting the velocity formula $v(t) = -32t + 96$ equal to 0 and solving for time t, you can determine *when* the velocity equals 0, namely at $t = 3$. Since you know the formula for position as a function of time, all you need to do is determine the height of the ball at time $t = 3$, the time it reaches its maximum height, by substituting 3 for t in the formula for $s(t)$:

$$s(3) = -16 \cdot 3^2 + 96 \cdot 3 = -144 + 288 = 144$$

Thus the ball goes up 144 feet before it starts coming down.

Example 2.14 Suppose a ball thrown up into the air has height $s(t) = -16t^2 + 144t$ feet after t seconds. Use Example 2.12 to determine how high the ball goes.

> *Solution:* In Example 2.12, the formula for the velocity of the ball was found to be $v(t) = -32t + 144$. As the ball reaches its high point at the instant when $v(t) = -32t + 144 = 0$, namely, at $t = 4.5$ seconds, you can see that the maximum height is
>
> $$s(4.5) = -16(4.5)^2 + 96(4.5) = -324 + 432 = 108 \text{ feet}$$

An even more fundamental application of derivatives is the following: Often one does not know the explicit formula for a quantity, and is interested in computing it. The information you do have about the quantity is not about the quantity itself, but about its rate of change. This means you know the derivative of the function, and want to find the function. Much of physics is like this. Newton's formulas give information about the acceleration of an object, that is, about the rate of change of velocity with respect to time. From this, one can often get information about the velocity, the rate of change of position with respect to time, and then information about the position itself of the object.

2.4 THE GEOMETRIC MEANING OF DERIVATIVES

Let (a, b) and (c, d) be two points in the plane, and consider the line segment connecting them. If this segment is neither horizontal nor vertical, it can be thought of as the hypotenuse of a right triangle, with the two short sides missing (see Figure 2.1). Fill in the two short sides, one horizontal and one vertical, and notice that the horizontal short side is the same length as the horizontal segment along the x-axis connecting a to c, while the vertical short side is the same length as the vertical segment along the y-axis connecting b to d.

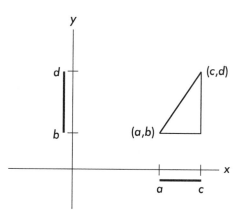

Figure 2.1. *The slope of a segment*

DEFINITION 2.3. *Let (a, b) and (c, d) be two points in the plane. The **slope** of the line segment connecting them is defined to be*

$$\frac{d - b}{c - a}$$

If $b = d$, so that the line segment is horizontal, then the slope is 0, while if $a = c$, so that the line segment is vertical, then the slope is undefined (as division by 0 is undefined).

Thus the slope of a line segment is the ratio formed by dividing the difference in the y-coordinates of two points on the line (represented by the vertical segment from b to d) by the difference in the x-coordinates (represented by the horizontal segment from a to c). When thought of like this, the slope should remind you of Equation 2.4.

Let $y = f(x)$ be a function, whose graph is pictured in Figure 2.2.

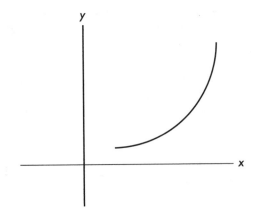

Figure 2.2. *A function*

Pick a general point x. Figure 2.3 shows x plotted on the x-axis, $f(x)$ plotted on the y-axis, and the point $(x, f(x))$ plotted on the graph of f.

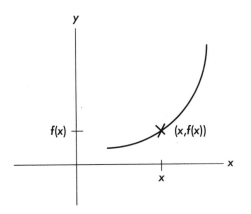

Figure 2.3. *A point on the graph*

Let h be a small positive number, so that $x + h$ is to the right of x on the number line. Figure 2.4 shows x and $x + h$ plotted on the x-axis, $f(x)$ and $f(x + h)$ plotted on the y-axis, and the points $(x, f(x))$ and $(x + h, f(x + h))$ plotted on the graph of f.

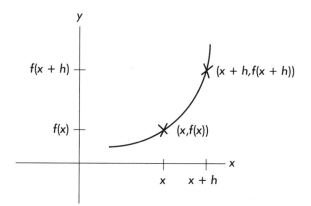

Figure 2.4. *Two points on the graph*

The change $\Delta x = (x+h) - x = h$ in the values of the variable is represented graphically by the length of the horizontal line segment from x to $x+h$ on the x-axis, while the corresponding change $\Delta y = f(x+h) - f(x)$ in the values of the function is represented graphically by the length of the vertical line segment from $f(x)$ to $f(x+h)$ on the y-axis, as pictured in Figure 2.5.

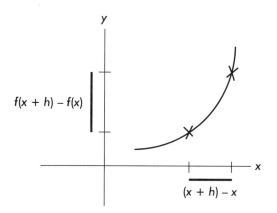

Figure 2.5. *Changes*

Connect the two points $(x, f(x))$ and $(x+h, f(x+h))$ on the graph of $y = f(x)$ with a straight line (see Figure 2.6). It is clear from Definition 2.3 that the slope of the straight line is

$$\frac{f(x+h) - f(x)}{(x+h) - x} = \frac{f(x+h) - f(x)}{h} = \frac{\Delta y}{\Delta x}$$

This is also the average rate of change of f with respect to x, as the variable changes from x to $x + h$.

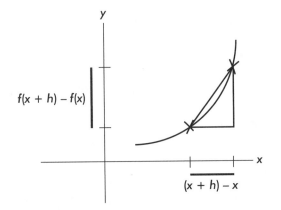

Figure 2.6. *Slope, or average rate of change*

What happens as h gets closer and closer to 0? Pick a smaller h and see. One unfortunate thing that happens is that the line segment connecting $(x, f(x))$ to $(x + h, f(x + h))$ on the graph gets shorter and becomes hard to see. So that you can see it better, Figure 2.7 shows several segments extended beyond the point $(x+h, f(x+h))$ on the graph. As h gets closer to 0, you get a succession of lines, all passing through $(x, f(x))$. Imagine getting one line from another by rotating or pivoting about the point $(x, f(x))$. As h gets closer and closer to 0, the lines in the picture seem to rotate or pivot so that they get closer and closer to what?

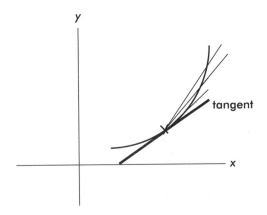

Figure 2.7. *Approaching the tangent line*

Hopefully, you can see that the lines get closer and closer to what you probably intuitively think of as the tangent line to the graph of f at the point $(x, f(x))$. As h gets closer and closer to 0, the lines pivot to become more closely aligned with the tangent line, and their slopes get closer and closer to the slope of the tangent line. As the slopes of the line segments are given by the formula $\dfrac{f(x+h) - f(x)}{h}$, it follows that the derivative of f at x,

$$f'(x) = \lim_{h \to 0} \frac{f(x+h) - f(x)}{h}$$

also gives the slope of the tangent line to the graph of f at the point $(x, f(x))$ on the graph.

Actually, the above considerations are used to give a precise definition of what exactly a tangent line is.

DEFINITION 2.4. *The **tangent line** to the graph of a function f at the point $(x, f(x))$ on the graph is the line passing through the point $(x, f(x))$ with slope equal to $f'(x)$.*

Let's revisit Example 2.14. A ball thrown up into the air has a height of $s(t) = -16t^2 + 144t$ feet after t seconds. How high does the ball go? Figure 2.8 shows the graph of the function, with t plotted on the horizontal axis, and s plotted on the vertical axis.

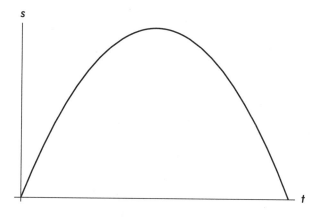

Figure 2.8. *The path of a ball*

Now which point on the graph corresponds to the high point of the ball? Can you stare at the figure, and see anything special *geometrically* about that point? Try drawing tangent lines. At the high point, the tangent line is horizontal and has slope 0. But slopes of tangent lines are given by derivatives, so the ball reaches its high point when $s'(t) = 0$. This is precisely the same conclusion as in Example 2.14, except there the interpretation of $s'(t)$ was as $v(t)$, the velocity of the ball at time t. However, you don't have to think about velocity. Just look at the graph of $s(t)$. The high point on the graph is the point where the tangent line has slope 0. To find that point, set the derivative equal to 0 and solve.

Example 2.15 Let $f(x) = x^2 + 4$. Using calculus, find the point where the graph has a horizontal tangent line.

Solution: Horizontal lines have slope 0, so you want to find the point on the graph where the tangent line has slope 0. But slopes of tangent lines are given by derivatives, so you want to solve $f'(x) = 0$ for x. Now

$$
\begin{aligned}
f(x+h) - f(x) &= \left((x+h)^2 + 4\right) - (x^2 + 4) \\
&= (x+h)^2 - x^2 \\
&= (x^2 + 2xh + h^2) - x^2 \\
&= 2xh + h^2
\end{aligned}
$$

This gives

$$
\begin{aligned}
f'(x) &= \lim_{h \to 0} \frac{f(x+h) - f(x)}{h} \\
&= \lim_{h \to 0} \frac{2xh + h^2}{h} \\
&= \lim_{h \to 0} (2x + h) \\
&= 2x + 0 = 2x
\end{aligned}
$$

Thus $f'(x) = 2x$, so that solving the equation $f'(x) = 0$ for x simply means solving the equation $2x = 0$ for x, to get $x = 0$. As $f(0) = 0^2 + 4 = 4$, $(0, 4)$ is the point on the graph of f where the tangent line is horizontal.

Example 2.16 Find the point on the graph of $f(x) = x^2 + 4$ where the tangent line has slope 4.

Solution: As the slope of the tangent line equals the derivative, you must solve $f'(x) = 4$. By Example 2.15, this means solving $f'(x) = 2x = 4$. Clearly $x = 2$ and $f(2) = 2^2 + 4 = 8$, so that $(2, 8)$ is the point on the graph where the tangent line has slope 4.

2.5 SUMMARY OF MAIN POINTS

- Calculus is a mathematical tool that allows the precise study of quantities that are continuously changing.

- A change in a quantity is determined by subtracting one value of the quantity from another value.

- A rate of change is a ratio, or fraction, with the change of one quantity in the numerator of the fraction, and the change in another quantity in the denominator of the fraction.

- A rate of change of quantity A with respect to quantity B always means that the change in A is in the numerator, while the change in B is in the denominator.

- Velocity is the rate of change of position with respect to time.

- An instantaneous rate of change is the limit of average rates of change, over smaller and smaller intervals.

- If y is a function of x, that is, if $y = f(x)$, then the derivative of f with respect to x is precisely the instantaneous rate of change in the values of f with respect to x.

- Standard notations for the derivative of $y = f(x)$ with respect to x are $f'(x), \dfrac{dy}{dx}, y'$, or $\dfrac{df}{dx}$.

- For $y = f(x)$,

$$f'(x) = \lim_{h \to 0} \frac{f(x+h) - f(x)}{h}$$

and

$$\frac{dy}{dx} = \lim_{\Delta x \to 0} \frac{\Delta y}{\Delta x}$$

Both notations mean the same thing.

- Depending upon the formula for $f(x)$, there are techniques for computing the derivative $f'(x)$.

- A geometric interpretation of $f'(x)$ is the slope of the tangent line to the graph of f at $(x, f(x))$.

- Setting the derivative $f'(x)$ equal to 0, and solving for x, is a common technique for finding both high and low points on the graph of a function $f(x)$.

2.6 EXERCISES

Do Exercises 1–8, first for the function $f(x) = x^2 + 3$, and then for the function $f(x) = x^2 - 4x + 3$.

1. Find the average rate of change of f with respect to x, as x changes from 2 to 5.

2. Find the average rate of change of f with respect to x, as x changes from 2 to 3.

3. Find the average rate of change of f with respect to x, as x changes from 2 to 2.1.

4. Find the average rate of change of f with respect to x, as x changes from 2 to 2.01.

5. Find the average rate of change of f with respect to x, as x changes from 2 to $2 + h$.

6. Find $f'(2)$, the *instantaneous* rate of change of f with respect to x, at $x = 2$.

7. Find the average rate of change of f with respect to x, as x changes from $x = a$ to $x = a + h$.

8. What happens to the answer to Exercise 7 as h gets closer and closer to 0? That is, what is $f'(a)$, the instantaneous rate of change of f with respect to x at $x = a$?

For Exercises 9–12, you can use the following formula for the derivative of a quadratic function: If $f(x) = ax^2 + bx + c$, then $f'(x) = 2ax + b$.

9. A ball thrown up into the air has height above the ground $s(t) = -16t^2 + 96t$ feet, t seconds after it is thrown.

   a. How long does it take the ball to reach its high point?
   b. How high does the ball go?
   c. At which times is the speed of the ball equal to 48 feet per second?
   d. With what velocity does the ball hit the ground?

10. A ball dropped from the roof of a building 240 feet tall has height above the ground $s(t) = -16t^2 + 240$ feet, t seconds after it is dropped.

   a. How long does it take the ball to hit the ground?
   b. With what velocity does the ball hit the ground?
   c. What is the average velocity of the ball for the first second after it is dropped?
   d. What is the instantaneous velocity of the ball at time $t = 1$?
   e. What is the average velocity of the ball between the time it is dropped and the time it hits the ground?

11. Let $f(x) = x^2 - 4x + 4$.

    a. At which point on the graph of f is the tangent line horizontal?

    b. At which point on the graph of f does the tangent line have slope 10?

    c. At which point on the graph of f does the tangent line have slope $\dfrac{1}{10}$?

12. Let $f(x) = 3x^2 - 6x + 4$.

    a. At which point on the graph of f is the slope of the tangent line to the graph equal to the average rate of change of f with respect to x between $x = 1$ and $x = 4$?

    b. At which point on the graph of f is the *instantaneous* rate of change of f with respect to x equal to the *average* rate of change of f with respect to x between $x = 0$ and $x = 3$?

    c. Solve the following equation for x: $f'(x) = 12$.

3
THE IDEA OF LIMITS

3.1 THE BASIC DEFINITION

In the previous chapter, the idea of **limits** came up naturally in the course of trying to define the derivative. In this chapter, you'll study limits more systematically.

Computing a limit just means computing what happens to the value of an expression as one of the variables in the expression gets closer and closer to (but does not equal) some particular value. Symbolically,

$$\lim_{x \to c} f(x)$$

stands for the phrase "the limit of $f(x)$, as x approaches c," and computing $\lim_{x \to c} f(x)$ just means computing what happens to the values of the function $f(x)$, if $f(x)$ is evaluated for values of x getting closer and closer to (but not equal to) the number c. If these values of $f(x)$ get closer and closer to some one particular number L, you say that

the limit of $f(x)$, as x approaches c, equals L

You write this as

$$\lim_{x \to c} f(x) = L$$

DEFINITION 3.1. *The expression* $\lim_{x \to c} f(x) = L$ *means that as x gets closer and closer to c, through values both smaller than c and larger than c, but not equal to c, then the values of $f(x)$ get closer and closer to the real number L.*

You should be aware that it may sometimes happen that there is no one particular number L that the values of $f(x)$ get closer and closer to as x approaches c, and in this case you say that "$\lim_{x \to c} f(x)$ **does not exist**." You'll get an idea of what can happen by looking at the following examples.

Example 3.1 Compute $\lim_{x \to 3} \dfrac{x^2 - 2}{2x + 1}$.

Solution: You want to determine what happens to the values of $\dfrac{x^2 - 2}{2x + 1}$ as you compute this expression for x-values getting closer and closer to 3, such as $x = 2.8$, 2.9, 2.99, 2.999 and

$x = 3.2, 3.1, 3.01, 3.001$. You want to see if the values of the expression get closer and closer to some particular number L. Check the values in the following tables yourself, with your calculator.

	x gets close to 3 from the left			
x	2.8	2.9	2.99	2.999
$\dfrac{x^2 - 2}{2x + 1}$	0.8848	0.9426	0.9943	0.9994

	x gets close to 3 from the right			
x	3.2	3.1	3.01	3.001
$\dfrac{x^2 - 2}{2x + 1}$	1.1135	1.0569	1.0057	1.0006

It certainly looks as though, when x is getting closer and closer to 3, whether from below (that is, from the left on the number line) or from above (that is, from the right on the number line), that the values of $\dfrac{x^2 - 2}{2x + 1}$ seem to be getting closer and closer to 1. Thus,

$$\lim_{x \to 3} \frac{x^2 - 2}{2x + 1} = 1$$

Look carefully at the above example again. Do you notice an easier way to do it? Suppose that, instead of calculating all the values in the table, you had simply substituted the value $x = 3$ into the expression. You would have found that

$$\frac{x^2 - 2}{2x + 1} = \frac{9 - 2}{6 + 1} = \frac{7}{7} = 1$$

Note that this is the same value obtained above for the limit. This seems like an easier way to compute limits. It certainly seems to make sense that as x gets closer and closer to 3, $x^2 = x \cdot x$ would get closer and closer to $3 \cdot 3 = 9$, and thus the numerator $x^2 - 2$ would get closer and closer to $9 - 2 = 7$. It also seems to make sense that as x gets closer and closer to 3, $2x$ would get closer and closer to $2 \cdot 3 = 6$, and thus the denominator $2x + 1$ would get closer and closer to $2 \cdot 3 + 1 = 7$. Finally, if the numerator gets closer and closer to 7, and the denominator gets closer and closer to 7, the whole expression obviously just gets closer and closer to $\dfrac{7}{7} = 1$.

The above reasoning actually justifies computing $\lim_{x \to 3} \dfrac{x^2 - 2}{2x + 1}$ by simply evaluating $\dfrac{x^2 - 2}{2x + 1}$ at $x = 3$. The method works because the operations of arithmetic, namely, addition, subtraction, multiplication, and division, all behave reasonably with respect to this idea of "getting closer and closer to" as long as nothing illegal happens. The one illegality you will mainly have to watch out for is the old cardinal sin of arithmetic—division by zero. You'll see examples of that later, but for now let's assume it won't come up.

I want to explain a little more carefully what it means to say that the operations of arithmetic behave reasonably with respect to the idea of "getting closer and closer to." It means simply that, for example, if x is getting closer and closer to 2 and y is getting closer and closer to 3, then $x + y$ is getting closer and closer to $2 + 3 = 5$, $x - y$ is getting closer and closer to $2 - 3 = -1$, $x \cdot y$ is getting closer and closer to $2 \cdot 3 = 6$, and $\dfrac{x}{y}$ is getting closer and closer to $\dfrac{2}{3} = 0.\bar{6}$. You can either believe all this because it seems so eminently reasonable and intuitively obvious, or convince yourself by playing around with a calculator or computer. In any case, this reasonableness of the operations of arithmetic justifies computing limits by **substituting in** the limiting value of the variable. The expression $\dfrac{x^2 - 2}{2x + 1}$ involves only the four basic operations of arithmetic, and the problem of division by zero does not come up, so computing $\lim_{x \to 3} \dfrac{x^2 - 2}{2x + 1}$ by just evaluating it at $x = 3$, that is, by just substituting the value $x = 3$, works.

Example 3.2 (a) Compute $\lim\limits_{x \to 4} \dfrac{2x + 9}{x - 3}$ by evaluating the expression at $x = 3.9$, 3.99, 3.999 and at $x = 4.1$, 4.01, 4.001.

(b) Compute $\lim\limits_{x \to 4} \dfrac{2x + 9}{x - 3}$ by substituting $x = 4$.

Solution: (a) Check for yourself with a calculator that the following tables of values are correct.

x gets close to 4 from the left			
x	3.9	3.99	3.999
$\dfrac{2x + 9}{x - 3}$	18.667	17.152	17.015

x gets close to 4 from the right			
x	4.1	4.01	4.001
$\dfrac{2x + 9}{x - 3}$	15.636	16.851	16.985

It certainly looks like the limit is 17.

(b) Substituting the value $x = 4$ yields

$$\frac{2 \cdot 4 + 9}{4 - 3} = \frac{17}{1} = 17$$

Thus, 17 is the limit. Notice that the problem of division by zero did not arise.

3.2 SOME COMPLICATIONS WITH THE DEFINITION OF LIMITS

Why bother at all with limits, then, in computing $\lim_{x \to c} f(x)$ when you can just substitute in the value $x = c$? One reason is that sometimes you have to deal with unusual functions. In a standard calculus course, these functions are introduced precisely for the purpose of emphasizing the *definition* of a limit, which really involves what happens to the values of $f(x)$ as x gets closer and closer to c, and not what the value of $f(x)$ at $x = c$ is. In this section you will see several examples of such functions. The second, and critically important reason for dealing with limits is that the most interesting limits generally arise *precisely* when substitution gives an illegal expression involving division by 0 (or even $\frac{0}{0}$). These types of limits always occur in the definition of the derivative, and will be dealt with in Sections 3.3 and 3.4.

A typical example of an unusual function is one defined by different formulas on different intervals of the line. The point of such examples is to emphasize that for a limit to exist, then, as x gets close to c by values smaller than c, and as x gets close to c by values larger than c, the values of $f(x)$ must get closer to the *same* number L.

Example 3.3 Let

$$f(x) = \begin{cases} x^2 - 4 & \text{for } x < 2 \\ x^2 & \text{for } x \geq 2 \end{cases}$$

Analyze $\lim_{x \to 2} f(x)$.

Solution: The fact that the formula changes at $x = 2$ should caution you not to simply evaluate $f(2)$. If you pick values of x getting closer and closer to 2 from below, such as $x = 1.9, 1.99, 1.999$, then you compute $f(x)$ by using $x^2 - 4$. As x gets closer and closer to 2, then $x^2 - 4$ gets closer and closer to $2^2 - 4 = 4 - 4 = 0$, which you could always check out with a calculator. If you pick values of x getting closer and closer to 2 from above, such as $x = 2.1, 2.01, 2.001$, then you compute $f(x)$ by using x^2. Now as x gets closer and closer to 2, x^2 gets closer and closer to $2^2 = 4$.

Thus, depending upon from which direction you approach 2, there are two different numbers that $f(x)$ gets closer and closer to, namely, 0 and 4. Since there is not one number L that $f(x)$ gets closer and closer to, no matter how x gets close to 2, the limit does not exist. (However, something called **one-sided limits** does exist. You'll see this covered later, on page 54.)

Example 3.4 Let

$$f(x) = \begin{cases} x^2 - 4 & \text{for } x \neq 2 \\ 7 & \text{for } x = 2 \end{cases}$$

Analyze $\lim_{x \to 2} f(x)$.

Solution: In other words, $f(x)$ is given by the formula $x^2 - 4$ for $x \neq 2$, while $f(2) = 7$. This example emphasizes the fact that the *definition* of the limit as $x \to 2$ really requires examining values of $f(x)$ as x gets close to, but does not equal, 2. For x not equal to 2, no matter whether x is larger than 2 or smaller than 2, you use the formula $x^2 - 4$ to compute $f(x)$. As x gets closer and closer to 2, then $x^2 - 4$ gets closer and closer to $2^2 - 4 = 4 - 4 = 0$. Thus

$$\lim_{x \to 2} f(x) = 0$$

This is true even though substituting $x = 2$ into f yields $f(2) = 7$. The limit as $x \to 2$ exists, but it is not equal to 7. It is equal to 0.

Example 3.5 Analyze $\lim_{x \to 3} \dfrac{x^2 - 6x + 9}{2x + 1}$.

Solution: This example does not involve a function with different formulas on different intervals of the line, as the above two example did, so you might just try substituting the value $x = 3$. Doing this gives

$$\frac{3^2 - 6 \cdot 3 + 9}{2 \cdot 3 + 1} = \frac{9 - 18 + 9}{6 + 1} = \frac{0}{7} = 0$$

The limit exists and equals 0. Just remember, division *by* zero is not allowed, but division of a nonzero number *into* zero makes perfectly good sense, and gives a value of zero.

Example 3.6 Analyze $\lim_{x \to 2} \dfrac{x^2 - 4x + 4}{x + 2}$.

Solution: Substituting $x = 2$ yields $\dfrac{4 - 8 + 4}{2 + 2} = \dfrac{0}{4} = 0$, so $\lim_{x \to 2} \dfrac{x^2 - 4x + 4}{x + 2} = 0$. You can also check this out with a calculator, by computing values of the expression for $x = 1.9,\ 1.99,\ 1.999$ and for $x = 2.1,\ 2.01,\ 2.001$.

3.3 THE PROBLEM OF DIVISION BY ZERO

Example 3.7 Analyze $\lim\limits_{x \to 3} \dfrac{2x + 1}{x^2 - 6x + 9}$.

Solution: If you try substituting $x = 3$ you get $\dfrac{7}{0}$, which is undefined. However, you can say something about the behavior of the values of $\dfrac{2x + 1}{x^2 - 6x + 9}$, as x approaches 3. The numerator is getting closer to 7, and the denominator is getting closer to 0. Check for yourself with a calculator that the following table gives correct values.

x	$2x + 1$	$x^2 - 6x + 9$	$\dfrac{2x + 1}{x^2 - 6x + 9}$
x gets close to 3 from the left			
2.9	6.8	0.01	680
2.99	6.98	0.0001	69,800
2.999	6.998	0.000001	6,998,000
x gets close to 3 from the right			
3.1	7.2	0.01	720
3.01	7.02	0.0001	70,200
3.001	7.002	0.000001	7,002,000

It certainly seems from the table that the values of the expression are getting very, very large as $x \to 3$. In fact, you might guess that if you continued the table, the values would get larger and larger. Indeed, you could get the values larger than any particular number you might specify in advance.

When this happens to the value of an expression, it is standard to say that the expression "is getting arbitrarily large" or is "going to infinity." The standard way of writing this, for Example 3.7 would be

$$\lim_{x \to 3} \frac{2x + 1}{x^2 - 6x + 9} = \infty$$

You could say that the limit equals ∞ (infinity), with the understanding that this is a slight abuse of language. As ∞ is not really a real number, the expression is not really getting closer and closer to one particular real number. Thus, technically, the limit does not exist. But because you do know something about the behavior of values of the expression, namely, that they are getting larger than any specified real number, saying that the limit equals infinity is a way to express this information.

Even in Example 3.7, the method of substituting the value $x = 3$ works, so that you do not really have to do calculations like those in the table. Substituting $x = 3$ gives $\frac{7}{0}$, which is undefined, but it does tell you that the numerator is getting closer and closer to 7 while the denominator is getting closer and closer to 0. Now you cannot divide by 0, but you can divide by numbers close to 0. To see clearly what is happening when you divide by numbers close to 0, suppose you are dividing by 0.0001. Written as a fraction, $0.0001 = \frac{1}{10,000}$. Now dividing by a fraction is the same as multiplying by its reciprocal. The reciprocal of $\frac{a}{b}$ is $\frac{b}{a}$. Thus dividing by $\frac{1}{10,000}$ is the same as multiplying by 10,000. Dividing by numbers close to 0 is the same as multiplying by numbers that are very large in magnitude. Thus, as x gets close to 3, the numerator is getting close to 7 and the denominator is getting close to 0, and you are actually multiplying numbers close to 7 by numbers that are very large in magnitude. Something close to 7, times something large, is really large (in fact, about 7 times as large). The expression $\frac{7}{0}$ can be interpreted in this way.

One further point about Example 3.7. The denominator $x^2 - 6x + 9 = (x - 3)^2$ is a perfect square, and is thus always greater than or equal to 0. As x gets close to 3, the numerator $2x + 1$ gets close to 7, which is also positive. Thus for x close to 3, $\frac{2x + 1}{x^2 - 6x + 9}$ is of the form $\frac{+}{+}$, which is always positive. As it is positive and getting arbitrarily large in magnitude or absolute value, then it is clearly getting arbitrarily large, and you can indeed say that the limit equals ∞.

Example 3.8 Analyze $\lim\limits_{x \to 3} \dfrac{2x + 1}{-x^2 + 6x - 9}$.

> *Solution:* You may recognize this as almost like Example 3.7, except that the sign of the denominator has been changed. A substitution of $x = 3$ yields $\frac{7}{0}$, which, as you have seen, can be interpreted as meaning that the expression gets arbitrarily large in

magnitude. However, since the numerator gets close to 7, and the denominator equals $-(x-3)^2$, which is always less than or equal to 0, the expression $\dfrac{2x+1}{-x^2+6x+9}$ is always of the form $\dfrac{+}{-}$ for x close to 3. Thus it is getting arbitrarily large in magnitude, but in the negative direction. In this instance, you can say that the limit equals $-\infty$.

One-sided Limits

There is one complication that did not appear in the two previous examples. In these examples, as x approached 3, whether from the left or the right on the number line, the sign of the denominator stayed the same ($+$ for Example 3.7 and $-$ for Example 3.8).

Example 3.9 Analyze $\displaystyle\lim_{x \to 3} \dfrac{2x+1}{x^2-9}$.

Solution: A substitution of $x = 3$ yields $\dfrac{7}{0}$, which, although undefined, can be correctly interpreted as meaning that the expression gets arbitrarily large in magnitude. But while the numerator is getting close to 7, and so will be positive, the denominator is getting close to 0, but could be positive or negative. In fact, for $x < 3, x^2 < 9$, the denominator will be negative, and $\dfrac{2x+1}{x^2-9}$ is of the form $\dfrac{+}{-} = -$. It remains negative while getting arbitrarily large in magnitude. For $x > 3, x^2 > 9$, the denominator will be positive, and $\dfrac{2x+1}{x^2-9}$ is of the form $\dfrac{+}{+} = +$. It remains positive while getting arbitrarily large in magnitude.

It follows that the expression $\dfrac{2x+1}{x^2-9}$ is not doing *one* particular thing as x gets close to 3, but *two* different things, depending upon whether x is getting close to 3 from the left or the right. Accordingly, there is not one particular thing you can say about $\lim_{x \to 3} \dfrac{2x+1}{x^2-9}$. However, it is often useful to be able to express the fact that some *one* thing is happening as x gets close to 3 from *one* side.

DEFINITION 3.2. *A **one-sided limit** expresses what happens to the values of an expression as one of the variables in the expression gets closer and closer to some particular value c from either the left on the number line (that is, through values less than c) or from the right on the number line (that is, through values greater than c.)*

If the expression is $f(x)$ and the variable x approaches c from the left, the notation is

$$\lim_{x \to c^-} f(x)$$

If x approaches c from the right, the notation is:

$$\lim_{x \to c^+} f(x)$$

In Example 3.9,

$$\lim_{x \to 3^+} \frac{2x + 1}{x^2 - 9} = \infty$$

while

$$\lim_{x \to 3^-} \frac{2x + 1}{x^2 - 9} = -\infty$$

If you go back and look at Example 3.3, you will see that even though the limit did not exist, $\lim_{x \to 2^-} f(x) = 0$, while $\lim_{x \to 2^+} f(x) = 4$.

Example 3.10 Analyze $\lim_{x \to 2} \dfrac{x + 2}{x^2 - 4x + 4}$.

> **Solution:** Substituting $x = 2$ yields $\dfrac{4}{4 - 8 + 4} = \dfrac{4}{0}$, which is unde-
> fined. However, the values of the expression are getting arbitrarily
> large in magnitude as x gets closer and closer to 2, and the only
> question is to analyze the sign of the expression. The numerator
> $x+2$ is getting closer and closer to 4, which is positive. The denom-
> inator equals $(x - 2)^2$, which is a perfect square, and is therefore
> always greater than or equal to 0. The expression thus has the
> form $\dfrac{+}{+} = +$ for x close to 2, and $\lim_{x \to 2} \dfrac{x + 2}{x^2 - 4x + 4} = \infty$.

Example 3.11 Analyze $\lim_{x \to -2} \dfrac{x^2 - 4x + 4}{x + 2}$.

> **Solution:** Substituting $x = -2$ yields $\dfrac{4 + 8 + 4}{0} = \dfrac{16}{0}$, which is
> undefined, but does tell you that the expression is getting ar-
> bitrarily large in magnitude. Clearly the numerator is positive
> for x close to -2. What about the denominator? Well, for $x <
> -2$, $x + 2 < -2 + 2 = 0$, so as x approaches -2 from the left, the
> denominator is negative, and the expression has the form $\dfrac{+}{-} = -$
> and gets arbitrarily large in the negative direction. However, for
> $x > -2$, $x + 2 > -2 + 2 = 0$, so as x approaches -2 from the
> right, the denominator is also positive, and the expression has the

form $\dfrac{+}{+} = +$ and gets arbitrarily large. As the expression does not do *one* particular thing as x approaches -2, the limit does not exist. However, the one-sided limits do exist:

$$\lim_{x \to -2+} \frac{x^2 - 4x + 4}{x + 2} = \infty$$

and

$$\lim_{x \to -2^-} \frac{x^2 - 4x + 4}{x + 2} = -\infty$$

3.4 THE CASE OF $\dfrac{0}{0}$

The most interesting and important situation with limits is that in which a substitution yields $\dfrac{0}{0}$. This is precisely the situation you'll be confronted with when attempting to compute derivatives from the definition. Suppose you have a fractional expression $\dfrac{n(x)}{d(x)}$, and as x is getting close to 3, both the numerator $n(x)$ and the denominator $d(x)$ are getting close to 0. Remember, dividing by $d(x)$ is like multiplying by its reciprocal $\dfrac{1}{d(x)}$, and as $d(x)$ gets close to 0, $\dfrac{1}{d(x)}$ gets arbitrarily large in magnitude. Thus, as x approaches 3, you are trying to study the behavior of $\dfrac{n(x)}{d(x)} = n(x) \cdot \dfrac{1}{d(x)}$, a product of two terms, one of which is getting arbitrarily small (close to 0), and the other of which is getting arbitrarily large. The situation is sort of like a race between the two terms, and the critical question is whether the term that's getting small is getting small faster, or about at the same rate, or slower than, the term that is getting large.

The examples that follow will show that just about anything can happen when a substitution yields $\dfrac{0}{0}$. Just substituting and getting $\dfrac{0}{0}$ does *not determine* the limit, and for this reason the situation is called **indeterminate**. Unlike the situation when a substitution yields, for example, $\dfrac{7}{0}$, which is *undefined* but can be interpreted, $\dfrac{0}{0}$ yields absolutely no information about the limit. It does not even tell you that the limit does not exist. The only thing it tells you is that you must do more work to determine the limit.

The following three expressions, which are almost transparently simple, hopefully will get the idea across. Consider the three functions $\dfrac{x}{x^2}$, $\dfrac{x}{x}$, and $\dfrac{x^2}{x}$. For now, don't simplify the functions using algebra. Just take them as given, to develop your intuition about limits. Let's compute $\lim_{x \to 0}$ for the three examples, using a calculator. Note that in all three cases, a substitution of $x = 0$ yields the *indeterminate* $\frac{0}{0}$. Check for yourself with a calculator that the following table is correct.

x	x^2	$\frac{x}{x^2}$	$\frac{x}{x}$	$\frac{x^2}{x}$
0.1	0.01	10	1	0.1
0.01	0.0001	100	1	0.01
0.001	0.000001	1000	1	0.001
0.0001	0.00000001	10,000	1	0.0001

It certainly looks from the numbers like $\lim_{x \to 0} \frac{x}{x^2} = \infty$. You can see from the table that as x goes to 0, x^2 goes to 0 even faster. Now dividing by x^2 is like multiplying by $\frac{1}{x^2}$, and as x^2 goes to 0, $\frac{1}{x^2}$ gets arbitrarily large. Since x^2 is going to 0 faster than x is (for example, when $x = 0.01 = \frac{1}{100}$, $x^2 = 0.0001 = \frac{1}{10,000}$), then $\frac{1}{x^2}$ is getting large faster than x is getting small. Thus, for $\frac{x}{x^2} = x \cdot \frac{1}{x^2}$, the term $\frac{1}{x^2}$ carries the day.

It also certainly looks from the table that $\lim_{x \to 0} \frac{x}{x} = 1$, and this is easily explainable. Certainly the numerator x and the denominator x are going to 0 at the same rate. The situation for the third function $\frac{x^2}{x}$ is the reverse of the first. The numerator x^2 is going to 0 faster than the denominator x is, so x^2 is going to 0 faster than $\frac{1}{x}$ is getting large. Thus, for $\frac{x^2}{x} = x^2 \cdot \frac{1}{x}$, the term x^2 carries the day. The table seems to confirm this analysis that $\lim_{x \to 0} \frac{x^2}{x} = 0$.

By now you've probably already noticed that all three functions can be simplified by canceling or factoring an x from the numerator and denominator. Thus $\frac{x}{x^2} = \frac{1}{x}$, $\frac{x}{x} = 1$, and $\frac{x^2}{x} = x$. Now the method of substituting the limiting value of $x = 0$ seems to work. For $\frac{1}{x}$ you get $\frac{1}{0}$, which, as you've seen in the last section, means that the expression is getting large in magnitude. (For this expression, you also have to worry about one-sided limits.) Substituting $x = 0$ to 1 just gives 1. For the second simplified function, the values don't really depend upon x, and thus stay the same (namely, 1), no matter what x is. Finally, substituting $x = 0$ in the third simplified function x clearly yields 0 as the limit.

The significance of what you have just done for these three simple functions is the following: by factoring an x from both the numerator and the denominator, you have canceled out precisely the term responsible for giving

$\frac{0}{0}$ in the original substitution of $x = 0$. After that cancellation, you can then substitute the limiting value of $x = 0$ in the simplified expression and get an answer that makes sense. This is the technique to try to follow in many limit problems where substitution yields $\frac{0}{0}$—try to rewrite the expression, using algebra, to cancel out from both the numerator and denominator that part of the expression responsible for giving the 0 in the numerator and denominator.

Example 3.12 Analyze $\lim\limits_{x \to 2} \dfrac{x^2 - 4}{x - 2}$.

Solution: First note that simply substituting the limiting value of $x = 2$ yields $\frac{0}{0}$, which is indeterminate. However, you get a 0 for the denominator because of the factor $x - 2$, evaluated at $x = 2$. If you can find a factor of $x - 2$ in the numerator, the factors can be canceled, eliminating the terms responsible for giving $\frac{0}{0}$. Perhaps for the simplified expression, substituting the value $x = 2$ will work. Now $x^2 - 4$ does factor as $(x - 2)(x + 2)$, so

$$\frac{x^2 - 4}{x - 2} = \frac{(x - 2)(x + 2)}{x - 2} = \frac{x + 2}{1} = x + 2$$

and clearly $\lim_{x \to 2} (x + 2) = 2 + 2 = 4$.

The above answer can be confirmed by checking the following table of values with a calculator.

x	$x^2 - 4$	$x - 2$	$\dfrac{x^2 - 4}{x - 2}$
x gets close to 2 from the left			
1.9	−0.39	−0.1	3.9
1.99	−0.0399	−0.01	3.99
1.999	−0.003999	−0.001	3.999
x gets close to 2 from the right			
2.1	0.41	0.1	4.1
2.01	0.0401	0.01	4.01
2.001	0.004001	0.001	4.001

You can see that the values of the expression seem to be getting closer and closer to 4, but wasn't the algebraic approach a lot easier?

Example 3.13 Analyze $\lim\limits_{x \to 2} \dfrac{x^2 - 4}{x^2 + x - 6}$.

Solution: Substitution of the limiting value $x = 2$ yields $\dfrac{0}{0}$. As the example involves polynomials in the numerator and the denominator, there should be a factor of $x - 2$ in each, which you can cancel. Thus

$$\frac{x^2 - 4}{x^2 + x - 6} = \frac{(x - 2)(x + 2)}{(x - 2)(x + 3)} = \frac{x + 2}{x + 3}$$

By substitution you now get

$$\lim_{x \to 2} \frac{x + 2}{x + 3} = \frac{4}{5} = 0.8$$

Limits and Derivatives

Recall from the previous chapter that the derivative $f'(x)$ of a function $f(x)$ is defined as

$$f'(x) = \lim_{h \to 0} \frac{f(x + h) - f(x)}{h}$$

If you try computing the limit by substituting the limiting value $h = 0$, you *always* get

$$\frac{f(x + 0) - f(x)}{0} = \frac{f(x) - f(x)}{0} = \frac{0}{0}$$

This is an indeterminate. Remember, this does not mean that the limit does not exist, just that more work must be done to determine whether or not the limit exists and what it is. What kind of extra work needs to be done? The previous examples point the way: the h in the denominator gives a value of 0 when you substitute the limiting value of $h = 0$, so you would clearly like to have a factor of h in the numerator to cancel the h in the denominator. Unfortunately, the numerator $f(x + h) - f(x)$ seems to have no factor of h in it. The idea behind the methods for computing the derivatives of certain types of functions directly from the definition of derivative is simply this—use algebraic techniques to rewrite $f(x + h) - f(x)$ so that it does have a factor of h in it. You can cancel this h with the h in the denominator, and then try substituting the limiting value $h = 0$. In the next chapter, you will learn about the different algebraic techniques that work for different types of functions.

3.5 CONTINUITY AND DIFFERENTIABILITY

In this section, you will find a brief, non-rigorous and intuitive explanation of two technical terms whose definitions involve limits.

Continuity

DEFINITION 3.3. *A function $f(x)$ is called **continuous at $x = c$** if, first of all, f is defined both at the point $x = c$ and on an interval (or half-interval) around $x = c$, and secondly, if $\lim_{x \to c} f(x) = f(c)$.*

What the definition means is that if you try to compute what happens to the values of $f(x)$ as x gets closer and closer to c, you can do this just by substituting in the value of f at $x = c$. In other words, the value of f at $x = c$ is just what it should be.

To get a better idea of what this means, look at the following example, where the value of f at $x = c$ is *not* what it should be, so that f is not continuous at $x = c$.

Example 3.14 Let:

$$f(x) = \begin{cases} x^2 & \text{for } x \neq 2 \\ 5 & \text{for } x = 2 \end{cases}$$

Is f continuous at $x = 2$?

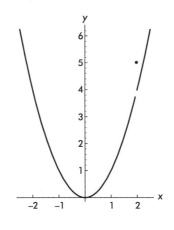

Figure 3.1

Solution: The formula for f makes sense for all numbers x, so f is defined on the whole real line. To compute $\lim_{x \to 2} f(x)$ you have to see what happens to the values of $f(x)$ as x gets close to, but does not equal, 2. For $x \neq 2$, the values of f are determined by the expression x^2, so clearly $\lim_{x \to 2} f(x) = \lim_{x \to 2} x^2 = 4$, while $f(2) = 5$. Thus f is not continuous at $x = 2$ because $\lim_{x \to 2} f(x) \neq f(2)$. If you look at the graph of f in Figure 3.1, you will see that the value of f at $x = 2$ is not "where it should be."

Another reason f may not be continuous at $x = c$ is that $\lim_{x \to c} f(x)$ may not exist. One common reason for this is that there may be *two* values, not *one*, that $f(x)$ gets closer and closer to as x gets closer and closer to c.

Example 3.15 Let

$$f(x) = \begin{cases} -1 & \text{for } x < 0 \\ 1 & \text{for } x \geq 0 \end{cases}$$

Is f continuous at $x = 0$?

Solution: The answer is no, because $\lim_{x \to 0} f(x)$ does not exist, although the two one-sided limits $\lim_{x \to 0+} f(x) = 1$ and $\lim_{x \to 0-} f(x) = -1$ both exist. The graph of f is pictured in Figure 3.2.

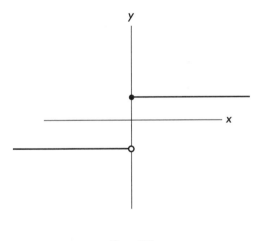

Figure 3.2

From both of the last two graphs, you can see intuitively that if f is defined at $x = c$, but not continuous at $x = c$, then there is a hole, or jump, or gap in the graph of f. If f is continuous at $x = c$, there is no such hole or gap.

DEFINITION 3.4. *a function f is said to be **continuous on an interval** if it is continuous at every point on the interval.*

This means that in the graph of f, there is no hole or gap or jump anywhere in the interval, and you can draw the graph (from one end of the interval to the other) without lifting your pencil off the paper. Try doing that for the above two *dis*continuous functions in the previous examples!!

Differentiability

Recall that the definition of derivative is in terms of a limit:

$$f'(x) = \lim_{h \to 0} \frac{f(x+h) - f(x)}{h}$$

DEFINITION 3.5. *The function f is called **differentiable at x = c** if the limit defining f'(c), namely,*

$$f'(c) = \lim_{h \to 0} \frac{f(c+h) - f(c)}{h} \tag{3.1}$$

exists.

As with continuity, you can get more insight into the meaning of the definition by looking at some cases where the definition does *not* hold.

One common reason the derivative may not exist is that as h approaches 0, the expression $\dfrac{f(c+h) - f(c)}{h}$ may get arbitrarily large. This means that the tangent line has slope ∞ (which is not a real number). Lines with undefined (or infinite) slope are precisely vertical lines, so in this case the tangent line to the graph is vertical. An example is pictured in Figure 3.3, for the function $f(x) = \sqrt{x}$, $x \geq 0$. The tangent line to the graph at the point $x = 0$ is vertical.

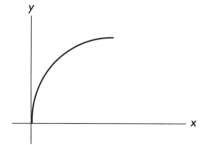

Figure 3.3. *The y-axis is a vertical tangent line at (0,0)*

As you've seen before, a second common reason a limit may not exist is that there may actually be two values that the expression gets closer and closer to, depending upon whether h gets close to 0 from the left or from the right. Consider the graph in Figure 3.4 of the function $f(x) = |x|$, at the point $c = 0$.

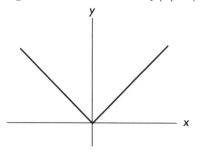

Figure 3.4. $y = |x|$

It is clear that for h positive,

$$\frac{f(0+h)-f(0)}{h} = \frac{h}{h} = 1$$

while for h negative, so that $|h| = -h$,

$$\frac{f(0+h)-f(0)}{h} = \frac{-h}{h} = -1$$

If you think of derivatives as giving slopes of tangent lines, you can also see what the problem is. What are reasonable possibilities for a tangent line to the above graph at $x = 0$? If you look at the left side of the graph, you might try the line with slope -1, and if you look at the right side of the graph, you might try the line with slope $+1$. In any case, there does not seem to be *one* obvious choice of a tangent line, so f is not differentiable at 0.

Geometrically, if f is differentiable at $x = c$, you could look at the graph of f and see that at $x = c$ there is a clear choice of a tangent line to the graph, and this tangent line will not be vertical. The example above, of $f(x) = |x|$, at $x = 0$, is typical of what happens when there is not *one* clear choice of tangent line—the graph has a sharp point. When f is differentiable at $x = c$, the graph will be **smooth** at the point $(c, f(c))$.

THEOREM 3.1. *If f is differentiable at $x = c$, then it is automatically continuous at $x = c$. The converse is not true—there are functions that are continuous at $x = c$ but not differentiable at $x = c$.*

Proof: See any standard calculus book for a proof of the first statement. An example of the second statement is $f(x) = |x|$ at $x = 0$. As mentioned, the graph has a sharp point at $x = 0$ so it is not differentiable there. However, you can draw its graph on an interval about 0 without lifting your pencil off the paper, so it is continuous at $x = 0$.

3.6 SUMMARY OF MAIN POINTS

- $\lim_{x \to c} f(x) = L$ means that the values of $f(x)$ get closer and closer to the real number L as x gets closer and closer to c.

- Some typical reasons $\lim_{x \to c} f(x)$ may not exist are the following:

  — The values of $f(x)$ get closer and closer to *two different numbers* as x gets close to c from the left, and as x gets close to c from the right. This situation is handled by the concept of *one-sided limits*.

  — The values of x get arbitrarily large (approach ∞) or arbitrarily small (that is, large in the negative direction, and approach $-\infty$). Notationally, this situation is usually described by saying that the limit is ∞ or $-\infty$, but *strictly speaking* the limit does not exist.

— The values of $f(x)$ *oscillate* or *fluctuate* as x approaches c. The graph in Figure 3.5 illustrates this behavior.

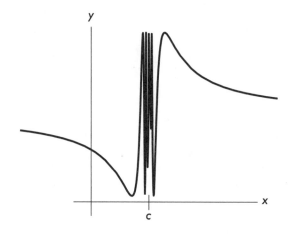

Figure 3.5. *An oscillating function*

- Limits behave as expected with respect to the operations of arithmetic, as long as division by zero is not involved. More technically, if $\lim_{x \to c} f(x) = A$ and $\lim_{x \to c} g(x) = B$, then

$$\lim_{x \to c} (f(x) + g(x)) = A + B$$
$$\lim_{x \to c} (f(x) - g(x)) = A - B$$
$$\lim_{x \to c} (f(x) \cdot g(x)) = A \cdot B$$
$$\lim_{x \to c} \frac{f(x)}{g(x)} = \frac{A}{B}, \text{ as long as } B \neq 0$$

The above rules often provide the justification for computing a limit by simply substituting the limiting value $x = c$.

- After substituting the limiting value of $x = c$, expressions such as

$$\frac{0}{A}, \text{ with } A \neq 0$$

can usually be interpreted as meaning the limit exists and equals 0. Expressions such as

$$\frac{A}{0}, \text{ with } A \neq 0$$

can usually be interpreted as meaning that the limit does not exist because the values converge (or diverge) to ∞. Expressions such as

$$\frac{0}{0}$$

are *indeterminate*. They cannot be interpreted without further work.

- Computing a derivative of $f(x)$ from the definition *always* involves computing an indeterminate limit. The techniques for handling this are often algebraic, and depend upon the formula for $f(x)$.

- A function $f(x)$ is said to be **continuous at c** if $\lim_{x \to c} f(x) = f(c)$. Geometrically, this means that the graph of f has no holes, jumps, or gaps at $x = c$.

- A function $f(x)$ is said to be **continuous on an interval** if it is continuous at every point of the interval. Geometrically, this means that the graph has no holes, jumps, or gaps anywhere in the interval. Thus, you can draw its graph from one end of the interval to the other without lifting your pencil off the paper.

- A function $f(x)$ is said to be **differentiable at x** if

$$\lim_{h \to 0} \frac{f(x + h) - f(x)}{h}$$

exists. Geometrically, this means that at the point $(x, f(x))$ on the graph, there is a well-defined tangent line (so the graph is *smooth* there, and does not have a sharp point), and furthermore that the tangent line is not vertical.

- It can happen that a function is continuous at a point, but not differentiable at that point. However, if a function is differentiable at a point, then it is automatically continuous at that point.

3.7 EXERCISES

In Exercises 1–14, either evaluate the given limit (as a real number or as ∞ or $-\infty$), or state that the limit does not exist. You should be able to do all of the problems without a calculator, but you can certainly check your answers on a calculator.

1. $\lim_{x \to 3} (x^2 - 4x + 9)$

2. $\lim_{x \to 2} \dfrac{x^2 + 4}{x + 8}$

3. $\lim_{x \to 1} \sqrt{7x + 9}$

4. $\lim_{x \to 0} [(x^2 + 4x + 3) \cdot (2x - 4)]$

5. $\lim_{x \to 3} \dfrac{x^2 - 9}{x^2 + 9}$

6. $\lim\limits_{x \to 0} \dfrac{3x}{3x + 7}$

7. $\lim\limits_{x \to 2} \dfrac{x^2}{x^2 - 4x + 4}$

8. $\lim\limits_{x \to 2+} \dfrac{x^2}{x^2 - 4x + 4}$

9. $\lim\limits_{x \to 2-} \dfrac{x^2}{x^2 - 4x + 4}$

10. $\lim\limits_{x \to 2} \dfrac{x^2}{x - 2}$

11. $\lim\limits_{x \to 2+} \dfrac{x^2}{x - 2}$

12. $\lim\limits_{x \to 2-} \dfrac{x^2}{x - 2}$

13. $\lim\limits_{x \to 3} \dfrac{x^2 - 9}{x - 3}$

14. $\lim\limits_{x \to 3} \dfrac{x^2 - 6x + 9}{x - 3}$

15. Let

$$f(x) = \begin{cases} 2x + 3 & \text{for } x < 1 \\ \\ 2x & \text{for } x \geq 1 \end{cases}$$

(a) Compute $\lim\limits_{x \to 1+} f(x)$.

(b) Compute $\lim\limits_{x \to 1-} f(x)$.

(c) Compute $\lim\limits_{x \to 1} f(x)$.

16. Let

$$f(x) = \begin{cases} 2x + 3 & \text{for } x < 1 \\ \\ 4 & \text{for } x \geq 1 \end{cases}$$

(a) Compute $\lim\limits_{x \to 1-} f(x)$.

(b) Compute $\lim\limits_{x \to 1+} f(x)$.

(c) Compute $\lim\limits_{x \to 1} f(x)$.

In the following problems, draw sketches of functions satisfying the given conditions. You do not have to find formulas for the functions.

17. A function that is not continuous at the point $x = 2$, but that is continuous at all other x's.

18. A function that is not continuous at either the point $x = 2$ or the point $x = 3$, but that is continuous at all other points.

19. A function that is continuous at $x = 2$ but is not differentiable at $x = 2$.

20. A different function that is continuous at $x = 2$ but is not differentiable at $x = 2$.

4
COMPUTING SOME DERIVATIVES

You can buy mathematical software, for less than $50, that will compute for you the derivatives of any functions you are ever likely to run across. So why should you learn how to compute derivatives for yourself? A reasonable answer is that you should know enough of what is going on to have an idea of what kinds of answers to expect, so that you can at least monitor that you are using the software and entering formulas correctly. If you have no idea of what to expect in the way of an answer, you will have nothing to clue you in to the fact that, for example, you might have made a typo entering the formula of the function to be differentiated. The reason for learning not only the *formulas* for computing derivatives, but also how to compute some derivatives *from the definition*, is to see that the rules and formulas for derivatives are not magical—they come from somewhere, and can be figured out and understood by real human beings. Education should not be a matter of memorizing mysterious formulas—it should be a matter of understanding.

Recall that for a function $f(x)$, the derivative $f'(x)$ is given by

$$f'(x) = \lim_{h \to 0} \frac{f(x + h) - f(x)}{h}$$

As always with the definition of the derivative, simply substituting the limiting value of $h = 0$ gives the indeterminate $\frac{0}{0}$. The general procedure for handling the indeterminate is to rewrite the numerator so that you can actually factor out an h, which will cancel with the h in the denominator. You can *then* try substituting the limiting value of $h = 0$.

There are three types of functions whose derivatives you'll learn how to compute in this chapter. Each type has its own algebraic technique for computing the derivative, and as long as you remember the technique that goes with each type, you should be OK. You will find as you read along in this chapter several quick reviews of algebra—if you do not need the reviews, then just skip over them.

The types of functions whose derivatives you will learn how to compute are those involving powers of x, fractional formulas with the variable x in the denominator, and square roots of x.

4.1 POWERS OF X

Let $f(x) = x^2$. Then

$$f'(x) = \lim_{h \to 0} \frac{(x+h)^2 - x^2}{h}$$

If you will look at Examples 2.9 and 2.12 in Section 2.2, "Instantaneous Rates of Change," and Example 2.13 in Section 2.3, "The Derivative," you'll see that the algebraic technique is to expand the square $(x+h)^2$.

Review of Algebra—Part 1

If you don't need this review, just skip it.

For any number b, and any positive integer n, b^n is defined as b multiplied by itself n times, that is

$$b^n = \underbrace{b \cdot b \cdots b}_{n \text{ times}}$$

Recall the commutative law of addition,

$$b + c = c + b$$

the commutative law of multiplication,

$$b \cdot c = c \cdot b$$

and the distributive law relating multiplication to addition:

$$a \cdot (b + c) = a \cdot b + a \cdot c = (b + c) \cdot a$$

From these laws, it follows that

$$
\begin{aligned}
(x+h)^2 &= (x+h) \cdot (x+h) \\
&= x \cdot (x+h) + h \cdot (x+h) && \text{(the distributive law)} \\
&= x \cdot x + x \cdot h + h \cdot x + h \cdot h && \text{(the distributive law again)} \\
&= x \cdot x + x \cdot h + x \cdot h + h \cdot h && \text{(the commutative law)} \\
&= x^2 + 2xh + h^2
\end{aligned}
$$

If you wanted to expand $(x+h)^3$, you could do the same thing:

$$
\begin{aligned}
(x+h)^3 &= (x+h) \cdot (x+h) \cdot (x+h) \\
&= (x+h)^2 \cdot (x+h) \\
&= (x^2 + 2xh + h^2) \cdot (x+h) \\
&= (x^2 + 2xh + h^2) \cdot x + (x^2 + 2xh + h^2) \cdot h && \text{(the distributive law)} \\
&= (x^2 \cdot x + 2xh \cdot x + h^2 \cdot x) \\
&\quad + (x^2 \cdot h + 2xh \cdot h + h^2 \cdot h) && \text{(the distributive law again)} \\
&= (x^3 + 2x^2h + h^2x) + (x^2h + 2xh^2 + h^3) && \text{(the commutative laws of} \\
&= x^3 + 3x^2h + 3xh^2 + h^3 && \text{addition and multiplication)}
\end{aligned}
$$

End of Algebra Review—Part 1

Back to the derivative of $f(x) = x^2$. Recall that

$$f'(x) = \lim_{h \to 0} \frac{(x+h)^2 - x^2}{h}$$

Expanding $(x+h)^2$ converts the numerator to

$$(x+h)^2 - x^2 = (x^2 + 2xh + h^2) - x^2$$

Now the x^2-terms cancel out, and all the remaining terms, $2xh$ and h^2, in the numerator have a factor of h in them. This can be factored out, to give

$$2xh + h^2 = (2x + h) \cdot h$$

The factor of h in the numerator cancels out the h in the denominator, and now the limit can be evaluated by substituting in the limiting value of $h = 0$, as follows:

$$
\begin{aligned}
f'(x) &= \lim_{h \to 0} \frac{(x+h)^2 - x^2}{h} \\
&= \lim_{h \to 0} \frac{(2x+h) \cdot h}{h} \\
&= \lim_{h \to 0} (2x + h) \\
&= 2x + 0 = 2x
\end{aligned}
$$

Notice that once you cancel out a factor of h in the numerator with the h in the denominator, you no longer obtain the indeterminate $\frac{0}{0}$ when substituting in the limiting value of $h = 0$. Thus,

$$\text{if } f(x) = x^2, \text{ then } f'(x) = 2x$$

Example 4.1 Find the derivative $f'(x)$ of $f(x) = x^3$.

Solution: Use the expansion of $(x+h)^3$ worked out in the algebra review on page 70.

$$
\begin{aligned}
f'(x) &= \lim_{h \to 0} \frac{(x+h)^3 - x^3}{h} \\
&= \lim_{h \to 0} \frac{(x^3 + 3x^2h + 3xh^2 + h^3) - x^3}{h} \\
&= \lim_{h \to 0} \frac{3x^2h + 3xh^2 + h^3}{h} \\
&= \lim_{h \to 0} \frac{(3x^2 + 3xh + h^2) \cdot h}{h} \\
&= \lim_{h \to 0} (3x^2 + 3xh + h^2) \\
&= 3x^2 + 3x \cdot 0 + 0^2 \\
&= 3x^2
\end{aligned}
$$

Notice that once you actually substitute the limiting value of $h = 0$, you should no longer write $\lim_{h \to 0}$. Once you substitute the limiting value, you have already taken the limit.

Example 4.2 Find the derivative $f'(x)$ of $f(x) = 4x^2$.

> **Solution:** This is handled much like x^2 was, but you must multiply by 4. Remember that the formula gives a recipe or procedure for computing a value from the x you give it. The procedure described by $f(x) = 4x^2$ is first to square what you are given, and then multiply by 4. Thus $f(x + h) = 4(x + h)^2$. Accordingly,
>
> $$
> \begin{aligned}
> f'(x) &= \lim_{h \to 0} \frac{4(x + h)^2 - 4x^2}{h} \\
> &= \lim_{h \to 0} \frac{4 \cdot (x^2 + 2xh + h^2) - 4x^2}{h} \\
> &= \lim_{h \to 0} \frac{(4x^2 + 8xh + 4h^2) - 4x^2}{h} \\
> &= \lim_{h \to 0} \frac{8xh + 4h^2}{h} \\
> &= \lim_{h \to 0} \frac{(8x + 4h) \cdot h}{h} \\
> &= \lim_{h \to 0} (8x + 4h) \\
> &= 8x + 4 \cdot 0 = 8x
> \end{aligned}
> $$

Example 4.3 Find the derivative $f'(x)$ of $f(x) = x^2 - 6x$.

> **Solution:** Since the function involves a power of x, you will have to expand powers. Remember as you do the algebra, that a minus sign distributes, so that $-(x + h) = -x - h$ and $-(x^2 - 6x) = -x^2 + 6x$. Thus,
>
> $$
> \begin{aligned}
> f'(x) &= \lim_{h \to 0} \frac{f(x + h) - f(x)}{h} \\
> &= \lim_{h \to 0} \frac{((x + h)^2 - 6(x + h)) - (x^2 - 6x)}{h} \\
> &= \lim_{h \to 0} \frac{(x^2 + 2xh + h^2 - 6x - 6h) - (x^2 - 6x)}{h} \\
> &= \lim_{h \to 0} \frac{2xh + h^2 - 6h}{h} \\
> &= \lim_{h \to 0} \frac{(2x + h - 6) \cdot h}{h}
> \end{aligned}
> $$

$$= \lim_{h \to 0} (2x + h - 6)$$
$$= 2x + 0 - 6 = 2x - 6$$

Notice that it is only *after* you actually divide out the h in the denominator with a factor of h in the numerator that you can substitute the limiting value of $h = 0$.

Remember that if a formula for f involves a power of x, then computing the derivative of f from the definition will involve expanding a power of $x + h$.

4.2 FRACTIONAL EXPRESSIONS WITH THE VARIABLE IN THE DENOMINATOR

Let $f(x) = \dfrac{1}{x}$. Then

$$f'(x) = \lim_{h \to 0} \frac{\frac{1}{x+h} - \frac{1}{x}}{h}$$

As always, simply substituting $h = 0$ will give you the indeterminate $\dfrac{0}{0}$, an indication that you must do more work. The algebraic technique for reworking the numerator so that it has a factor of h is probably not too surprising to you if you have some experience in manipulating fractions—use common denominators.

Review of Algebra—Part 2

As before, if you do not need this review, just skip it.

For any number $c \neq 0$, dividing by c is like multiplying by the reciprocal $\dfrac{1}{c}$ of c, so that

$$\frac{a}{c} = a \cdot \frac{1}{c}$$

Furthermore, one way of thinking about what $\dfrac{1}{c}$ means is that it is the number that, when multiplied by c, gives 1, so that

$$c \cdot \frac{1}{c} = \frac{c}{c} = 1$$

From these simple facts, and the commutative and distributive laws, you can derive most of the rules for manipulating fractions.

RULE 1: If b and d are nonzero, then

$$\frac{1}{b} \cdot \frac{1}{d} = \frac{1}{bd}$$

Why? As $\dfrac{1}{bd}$ is the number that, when multiplied by bd, gives 1, to check the rule you need only check that $\dfrac{1}{b} \cdot \dfrac{1}{d}$ satisfies this property, that is, that

$$\frac{1}{b} \cdot \frac{1}{d} \cdot b \cdot d = 1$$

Simplifying gives

$$\frac{1}{b} \cdot \frac{1}{d} \cdot b \cdot d \;=\; \frac{1}{b} \cdot b \cdot \frac{1}{d} \cdot d \quad \text{by commutativity of multiplication}$$
$$= 1 \cdot 1 = 1$$

RULE 2: To multiply two fractions, you multiply the numerators and multiply the denominators:

$$\frac{a}{b} \cdot \frac{c}{d} = \frac{ac}{bd}$$

This is because

$$\frac{a}{b} \cdot \frac{c}{d} \;=\; a \cdot \frac{1}{b} \cdot c \cdot \frac{1}{d}$$
$$= a \cdot c \cdot \frac{1}{b} \cdot \frac{1}{d}$$
$$= a \cdot c \cdot \frac{1}{bd}$$
$$= \frac{ac}{bd}$$

RULE 3: To add two fractional expressions with the same denominator, simply add the numerators:

$$\frac{a}{b} + \frac{c}{b} = \frac{a+c}{b}$$

This is because

$$\frac{a}{b} + \frac{c}{b} \;=\; a \cdot \frac{1}{b} + c \cdot \frac{1}{b}$$
$$= (a+c) \cdot \frac{1}{b} \quad \text{(the distributive law)}$$
$$= \frac{a+c}{b}$$

RULE 4: To cancel common factors, use the formula

$$\frac{ac}{bc} = \frac{a}{b}$$

This is because, by Rule 2,

$$\frac{ac}{bc} = \frac{a}{b} \cdot \frac{c}{c}$$
$$= \frac{a}{b} \cdot 1$$
$$= \frac{a}{b}$$

RULE 5: To add fractions with different denominators, express both fractions with the same **common denominator**:

$$\frac{a}{b} + \frac{c}{d} = \frac{ad}{bd} + \frac{bc}{bd} \quad \text{(by Rule 4)}$$
$$= \frac{ad + bc}{bd} \quad \text{(by Rule 3)}$$

You get the denominator of the sum by multiplying denominators, and you get the numerator of the sum by cross-multiplying numerators and denominators, as follows in Figure 4.1:

Figure 4.1. *Common denominators*

You operate similarly for subtraction:

$$\frac{a}{b} - \frac{c}{d} = \frac{ad - bc}{bd}$$

The only part of the review left is division of fractions. Here there is one item you must be careful about, which didn't occur for addition and multiplication of fractions. We have used, without mentioning it, the associative laws of addition and multiplication, namely, that $a + (b + c) = (a + b) + c$ and $a(bc) = (ab)(c)$. What these laws imply is that if you want to add, or multiply, three numbers together, you can pair them any way you want, and you do not even have to put in parentheses, so that expressions such as $a + b + c$ or abc make sense. The expression $\dfrac{\frac{a}{b}}{c}$ makes no sense, because if you first divide a by b, and then divide this result by c, you can get a different answer than if you divide a by $\dfrac{b}{c}$. As an example, let $a = 24, b = 6$, and $c = 2$. Then

$$\frac{\left(\frac{24}{6}\right)}{2} = \frac{4}{2} = 2, \text{ but } \frac{24}{\left(\frac{6}{2}\right)} = \frac{24}{3} = 8$$

Thus the expression $\dfrac{24}{\frac{6}{2}}$ makes no sense, because it matters in which order the numbers are paired. One way of specifying the pairing is by doing what I just did, namely, using parentheses. Remember the old rule—do what is in the parentheses first. A less cluttered way of doing this is to use different-size horizontal lines, with the biggest line separating the main numerator from the main denominator. Thus,

$$\frac{24}{\frac{6}{2}}$$

means that 24 is the numerator, and $\frac{6}{2} = 3$ is the denominator, while

$$\frac{\frac{24}{6}}{2}$$

indicates that $\frac{24}{6} = 4$ is the numerator, while 2 is the denominator.

In working with expressions like this, be careful drawing your horizontal lines—at the moment you are writing something you may know what the main numerator and main denominator are, but if you continue on with your work and then have to go back to it later, you might have forgotten if it is not indicated clearly.

RULE 6: To divide fractional expressions, use the formula

$$\frac{\frac{a}{b}}{\frac{c}{d}} = \frac{ad}{bc}$$

If the numerator or denominator is not a fractional expression, you can still use this rule by writing any quantity x as $\dfrac{x}{1}$. To explain why this rule holds, first look at a special case:

$$\frac{1}{\frac{c}{d}} = \frac{d}{c}$$

This holds because by definition $\dfrac{1}{\frac{c}{d}}$ is just the number that, when multiplied by $\dfrac{c}{d}$, gives 1. To see that $\dfrac{d}{c}$ is that number, notice that

$$\frac{c}{d} \cdot \frac{d}{c} = \frac{cd}{dc} = \frac{cd}{cd} = 1$$

The general rule follows because

$$\begin{aligned}
\frac{\frac{a}{b}}{\frac{c}{d}} &= \frac{a}{b} \cdot \frac{1}{\frac{c}{d}} \\
&= \frac{a}{b} \cdot \frac{d}{c} \\
&= \frac{ad}{bc}
\end{aligned}$$

Just remember that dividing by a quantity $\dfrac{c}{d}$ is like multiplying by its reciprocal $\dfrac{1}{\frac{c}{d}} = \dfrac{d}{c}$.

End of Algebra Review—Part 2

Now, back to the derivative of $f(x) = \dfrac{1}{x}$. The derivative is

$$
\begin{aligned}
f'(x) &= \lim_{h \to 0} \frac{\frac{1}{x+h} - \frac{1}{x}}{h} \\[2ex]
&= \lim_{h \to 0} \frac{\frac{1 \cdot x - (x+h) \cdot 1}{(x+h)x}}{h} \qquad \text{(by applying Rule 5 to the numerator)} \\[2ex]
&= \lim_{h \to 0} \frac{\frac{x - (x+h)}{(x+h)x}}{h} \\[2ex]
&= \lim_{h \to 0} \frac{x - x - h}{(x+h)x} \cdot \frac{1}{h} \\[2ex]
&= \lim_{h \to 0} \frac{-h}{(x+h)x} \cdot \frac{1}{h} \\[2ex]
&= \lim_{h \to 0} \frac{-h}{(x+h)xh} \\[2ex]
&= \lim_{h \to 0} \frac{-1}{(x+h)x} \\[2ex]
&= \frac{-1}{(x+0)x} \\[2ex]
&= \frac{-1}{x^2}
\end{aligned}
$$

Note that you must be careful about the use of parentheses. When you write $\dfrac{1}{x+h} - \dfrac{1}{x}$ in common denominator form, the numerator is found by taking x and *subtracting* the quantity $x + h$. This *must* be written with parentheses, as $x - (x + h) = x - x - h$. If you don't use the parentheses, you will not properly distribute the minus sign, and you will get the wrong answer. The second thing to notice is that as before, you do not substitute the limiting value of $h = 0$ until you have actually divided out the h in the denominator with the factor of h in the numerator. Finally, when you have substituted the limiting value $h = 0$, you no longer write $\lim_{h \to 0}$.

Remember, to find the derivatives of functions in fractional form with the variable in the denominator, use common denominators. Also, be careful about the proper use of parentheses and about distributing minus signs.

In the following examples, some of the steps done above will be skipped.

Example 4.4 Find the derivative of $f(x) = \dfrac{3}{x}$.

Solution:

$$f'(x) = \lim_{h \to 0} \frac{\frac{3}{x+h} - \frac{3}{x}}{h}$$

$$= \lim_{h \to 0} \frac{\frac{3x - 3(x+h)}{(x+h)x}}{h}$$

$$= \lim_{h \to 0} \frac{3x - 3x - 3h}{(x+h)xh}$$

$$= \lim_{h \to 0} \frac{-3h}{(x+h)xh}$$

$$= \lim_{h \to 0} \frac{-3}{(x+h)x}$$

$$= \frac{-3}{(x+0)x}$$

$$= \frac{-3}{x^2}$$

Example 4.5 Find the derivative of $g(t) = \dfrac{1}{7t}$.

Solution: Note that changing the name of the letter you use for the variable and changing the name of the letter you use for the function has no effect at all on the work you do to find the derivative. The biggest difference between this example and the one before is that here the constant 7 appears in the denominator, and in the previous example, the constant 3 appeared in the numerator.

$$g'(t) = \lim_{h \to 0} \frac{g(t+h) - g(t)}{h}$$

$$= \lim_{h \to 0} \frac{\frac{1}{7(t+h)} - \frac{1}{7t}}{h}$$

$$= \lim_{h \to 0} \frac{7t - 7(t+h)}{7(t+h) \cdot 7t \cdot h}$$

$$= \lim_{h \to 0} \frac{7t - 7t - 7h}{7(t+h) \cdot 7t \cdot h}$$

$$= \lim_{h \to 0} \frac{-7}{7(t+h) \cdot 7t}$$

$$= \frac{-7}{7(t+0) \cdot 7t}$$

$$= \frac{-1}{7t^2}$$

4.3 SQUARE ROOTS

Let $f(x) = \sqrt{x}$, so that

$$f'(x) = \lim_{h \to 0} \frac{\sqrt{x+h} - \sqrt{x}}{h}$$

The technique here for rewriting the numerator so that a factor of h appears is called **rationalizing the numerator**. The numerator involves two different square roots, and you can get rid of the roots by multiplying by the same two square roots, but with the sign between them changed. So here you change the sign from a minus sign between the square roots to a plus sign between the square roots, and multiply by

$$\sqrt{x+h} + \sqrt{x}$$

Of course, you are not really allowed to change the expression $\sqrt{x+h} - \sqrt{x}$ you are given. You are only allowed to change the way you write it. The only thing you can multiply an expression by without changing it is 1, so multiply by 1, but a complicated 1, namely

$$1 = \frac{\sqrt{x+h} + \sqrt{x}}{\sqrt{x+h} + \sqrt{x}}$$

Remember that for any nonzero quantity c, $\dfrac{c}{c} = 1$. Now we can rewrite the original fraction as

$$\frac{\sqrt{x+h} - \sqrt{x}}{h} = \frac{\sqrt{x+h} - \sqrt{x}}{h} \cdot \frac{\sqrt{x+h} + \sqrt{x}}{\sqrt{x+h} + \sqrt{x}} \tag{4.1}$$

The two fractional expressions can be multiplied together by multiplying the numerators and multiplying the denominators. If you multiply the numerators on the right-hand side of Equation 4.1 you get

$$\left(\sqrt{x+h} - \sqrt{x}\right) \cdot \left(\sqrt{x+h} + \sqrt{x}\right)$$
$$= \sqrt{x+h}\sqrt{x+h} - \sqrt{x}\sqrt{x+h} + \sqrt{x+h}\sqrt{x} - \sqrt{x}\sqrt{x}$$

Now the middle two terms on the right-hand side of the equation are the same, except they appear with opposite signs, so they cancel (which is precisely why the sign between the square root terms was changed). Also, for any positive c, $\sqrt{c}\sqrt{c} = c$, as the square root of c is precisely the positive number that when squared gives c. It follows that the right-hand side of Equation 4.1 reduces to

$$(x + h) - x = h$$

This certainly has a factor of h in it. Thus,

$$
\begin{aligned}
f'(x) &= \lim_{h \to 0} \frac{\sqrt{x+h} - \sqrt{x}}{h} \\
&= \lim_{h \to 0} \frac{\sqrt{x+h} - \sqrt{x}}{h} \cdot \frac{\sqrt{x+h} + \sqrt{x}}{\sqrt{x+h} + \sqrt{x}} \\
&= \lim_{h \to 0} \frac{h}{h \cdot \left(\sqrt{x+h} + \sqrt{x}\right)} \\
&= \lim_{h \to 0} \frac{1}{\sqrt{x+h} + \sqrt{x}} \\
&= \frac{1}{\sqrt{x+0} + \sqrt{x}} \\
&= \frac{1}{2\sqrt{x}}
\end{aligned}
$$

Remember that for functions involving roots, the technique for finding the derivative from the definition is *rationalizing*, which is done by multiplying the two different square roots by a fractional expression with identical numerator and denominator, namely, the same two square roots with the sign between them changed.

4.4 SOME HARDER EXAMPLES

You've studied the different techniques for computing the derivative from the definition for functions involving either powers, fractional expressions, or roots. You might naturally expect that if the formula for a function involves several of these elements, you would have to combine techniques. That is correct, as the following examples show.

Example 4.6 Find the derivative $f'(x)$ of $f(x) = \dfrac{1}{\sqrt{x}}$.

Solution: As the formula for f involves both fractional expressions and square roots, you will have to use both common denominators and rationalization.

$$
\begin{aligned}
f'(x) &= \lim_{h \to 0} \frac{\frac{1}{\sqrt{x+h}} - \frac{1}{\sqrt{x}}}{h} \\
&= \lim_{h \to 0} \frac{\sqrt{x} - \sqrt{x+h}}{h\sqrt{x+h}\sqrt{x}} \quad\quad \text{(common denominator)} \\
&= \lim_{h \to 0} \frac{\sqrt{x} - \sqrt{x+h}}{h\sqrt{x+h}\sqrt{x}} \cdot \frac{\sqrt{x} + \sqrt{x+h}}{\sqrt{x} + \sqrt{x+h}} \quad\quad \text{(rationalizing)}
\end{aligned}
$$

$$= \lim_{h \to 0} \frac{x - (x + h)}{h\sqrt{x + h}\sqrt{x} \cdot \left(\sqrt{x} + \sqrt{x + h}\right)}$$

$$= \lim_{h \to 0} \frac{-h}{h\sqrt{x + h}\sqrt{x} \cdot \left(\sqrt{x} + \sqrt{x + h}\right)}$$

$$= \lim_{h \to 0} \frac{-1}{1 \cdot \sqrt{x + h}\sqrt{x} \cdot \left(\sqrt{x} + \sqrt{x + h}\right)}$$

$$= \frac{-1}{\sqrt{x}\sqrt{x}(\sqrt{x} + \sqrt{x})}$$

$$= \frac{-1}{2x\sqrt{x}}$$

Example 4.7 Let $f(x) = \dfrac{1}{x^2}$. Find $f'(x)$.

Solution: As the formula for f involves both fractional expressions and powers, you should expect both to have to use common denominators and to expand the power.

$$\begin{aligned}
f'(x) &= \lim_{h \to 0} \frac{\frac{1}{(x+h)^2} - \frac{1}{x^2}}{h} \\[2mm]
&= \lim_{h \to 0} \frac{x^2 - (x + h)^2}{h(x + h)^2 x^2} && \text{(common denominator)} \\[2mm]
&= \lim_{h \to 0} \frac{x^2 - (x^2 + 2xh + h^2)}{h(x + h)^2 x^2} && \text{(expand the power)} \\[2mm]
&= \lim_{h \to 0} \frac{-2xh - h^2}{h(x + h)^2 x^2} && \text{(cancel terms)} \\[2mm]
&= \lim_{h \to 0} \frac{(-2x - h) \cdot h}{h(x + h)^2 x^2} \\[2mm]
&= \lim_{h \to 0} \frac{-2x - h}{(x + h)^2 x^2} \\[2mm]
&= \frac{-2x}{x^2 \cdot x^2} \\[2mm]
&= \frac{-2}{x^3}
\end{aligned}$$

You may have wondered why we did not expand $(x + h)^2$ in the denominator also. The reason is that it is only the numerator that must be worked on so that a factor of h appears. Expanding the square in the denominator would not be wrong, it would just involve unnecessary extra work.

4.5 SUMMARY OF MAIN POINTS

- To compute the derivative

$$f'(x) = \lim_{h \to 0} \frac{f(x+h) - f(x)}{h}$$

  from the definition, some algebraic techniques must be used. The particular technique to be used depends upon the particular type of formula for $f(x)$.

- These algebraic techniques serve the purpose of factoring out the term h from the numerator $f(x+h) - f(x)$, so that you can cancel this h with the h in the denominator, and then substitute the limiting value of $h = 0$.

- For functions involving powers of x, the technique is to expand the corresponding power of $x + h$.

- For functions involving fractional expressions with the variable in the denominator, the technique involves rewriting $f(x+h) - f(x)$ using common denominators.

- For functions involving square roots, the technique is that of rationalizing the numerator.

- For functions involving several of the above types of formulas, the appropriate techniques must be combined.

4.6 EXERCISES

Compute the derivatives of the following functions, using *only* the *definition* of the derivative.

1. $f(x) = 2x^3$

2. $f(x) = 3x^2 + 4x - 2$

3. $f(x) = x^3 - 6x^2 + 3$

4. $h(t) = \dfrac{3}{7t}$

5. $p(x) = \dfrac{3x+1}{7x-6}$

6. $g(t) = \dfrac{1}{t^2 + 1}$

7. $h(u) = \sqrt{3u}$

8. $p(t) = \sqrt{4t - 2}$

9. $f(x) = \sqrt{x^2 + 1}$

10. $f(x) = \dfrac{1}{\sqrt{3x - 4}}$

5
FORMULAS FOR DERIVATIVES

If you thought that the last two examples of the previous chapter were difficult, then you'll be glad to learn that there are formulas you can use to compute derivatives without resorting to the definition of the derivative each time. Of course, the validity of these formulas is based upon the definition of the derivative, along with facts about limits. While I will give sketchy verifications of some of these formulas, I will not give precise proofs. You can find these in just about any standard calculus text. What I am more interested in doing is emphasizing to you that you should not just memorize formulas with symbols, but rather you should memorize the formulas in terms of *words*. For example, if you memorized the formula for the derivative of a quotient $\frac{f(x)}{g(x)}$ in terms of the symbols f and g, then you are likely to get confused if you have to compute the derivative of $\frac{x^2 + x}{x^3 - x^2 - 4}$, or worse, the derivative of $\frac{a(x)}{b(x)}$, or *even worse* the derivative of $\frac{g(x)}{f(x)}$. However, if you memorize the procedure for finding the derivative of a quotient in terms of words, you can repeat it to yourself to guide you in computing the derivative of any quotient function. Each derivative formula will be given both symbolically and in terms of words. So that you can get used to both the $f'(x)$ and the $\frac{dy}{dx}$ notation for derivatives, each formula will be given in terms of both notations, and different examples will be worked out using the various notations. For example, the problem of finding the derivative of the function $x^2 + x^3$ could start out as

$$\frac{d(x^2 + x^3)}{dx} = \ldots, \text{ or } \frac{d}{dx}(x^2 + x^3) = \ldots$$

or, using the *prime* notation, as

$$(x^2 + x^3)' = \ldots$$

5.1 DERIVATIVES OF SOME PARTICULAR FUNCTIONS

THEOREM 5.1.

1. If $f(x) = c$, a constant, then $f'(x) = 0$.

2. $\dfrac{dc}{dx} = 0$

3. The derivative of a constant is zero.

If you think of a derivative as measuring a rate of change, then since a constant does not change, it has a zero rate of change, and thus a derivative of 0. If you think of a derivative as measuring slopes of tangent lines, then since a constant function has a horizontal straight line as a graph, and horizontal lines have zero slope, the slope of the tangent line is 0 (as straight lines are their own tangents). Finally, if $f(x) = c$, a constant, then $f(x+h) = c$ also, for all h, so that

$$
\begin{aligned}
f'(x) &= \lim_{h \to 0} \frac{f(x+h) - f(x)}{h} \\
&= \lim_{h \to 0} \frac{c - c}{h} \\
&= \lim_{h \to 0} \frac{0}{h} \\
&= \lim_{h \to 0} 0 \\
&= 0
\end{aligned}
$$

THEOREM 5.2. *The Power Rule*

1. If $f(x) = x^n$, then $f'(x) = nx^{n-1}$.

2. $\dfrac{d(x^n)}{dx} = nx^{n-1}$

3. To take the derivative of x raised to a power, you multiply in front by the exponent, and subtract 1 from the exponent.

The power rule will not be proven. Notice, however, that in the case when n is a natural number, then the technique of expanding the power serves to prove the power rule, just as was done in Section 4.1.

Example 5.1 Use the power rule to find the derivative of $f(x) = x^2$.

Solution: $\dfrac{d(x^2)}{dx} = 2x^{2-1} = 2x^1 = 2x$. Compare this with the work in Section 4.1.

Example 5.2 Use the power rule to find the derivative of $f(x) = x^3$.

Solution: $(x^3)' = 3(x^{3-1}) = 3x^2$. Compare this with Example 4.1.

Example 5.3 Find $g'(t)$ for $g(t) = t^{13}$.

Solution: As you have seen before, changing the name of the variable and changing the name of the function have no effect on the procedure for finding derivatives. The function g is a power of the variable t, so just multiply in front by the exponent and then subtract 1 from the exponent to obtain $g'(t) = \dfrac{d(t^{13})}{dt} = 13t^{12}$.

For the next example, recall that for any nonzero real number c, $c^0 = 1$.

Example 5.4 Find the derivative $f'(x)$ of $f(x) = x$.

Solution: This derivative is almost trivial to find from the definition of derivatives, but the power rule works here also. As $f(x) = x = x^1$, it follows that $f'(x) = 1 \cdot x^0 = 1 \cdot 1 = 1$.

The remarkable fact about the power rule is that it works not only for exponents that are natural numbers, but for *any* exponent. Recall from algebra that $\sqrt{c} = c^{\frac{1}{2}}$, so that square roots can be changed to exponents, and that $\dfrac{1}{c^n} = c^{-n}$, so that sometimes fractional expressions can be changed to exponent form. Recall also that when you multiply factors with the same base then the exponents add, so that $c^n c^m = c^{n+m}$, and that when you divide factors with the same base, the exponents subtract, so that $\dfrac{c^n}{c^m} = c^{n-m}$.

When you compute the derivative of a function with the $\sqrt{}$ sign in it, always change the $\sqrt{}$ sign to exponent form, so that you can take advantage of the power rule.

Example 5.5 Use the power rule to compute the derivative $f'(x)$ of $f(x) = \sqrt{x}$.

Solution: As $f(x) = \sqrt{x} = x^{\frac{1}{2}}$, the power rule applies to give $f'(x) = \dfrac{1}{2}x^{(\frac{1}{2}-1)} = \dfrac{1}{2}x^{\frac{-1}{2}}$. If you use the rules for exponents to rewrite the answer, you see that $\dfrac{1}{2}x^{\frac{-1}{2}} = \dfrac{1}{2}\left(\dfrac{1}{x^{\frac{1}{2}}}\right) = \dfrac{1}{2}\left(\dfrac{1}{\sqrt{x}}\right) = \dfrac{1}{2\sqrt{x}}$, the same answer obtained in Section 4.3.

Example 5.6 Find the derivative of $f(x) = \dfrac{1}{\sqrt{x}}$.

> *Solution:* As $f(x) = \dfrac{1}{\sqrt{x}} = \dfrac{1}{x^{\frac{1}{2}}} = x^{\frac{-1}{2}}$, by the power rule $f'(x) = \dfrac{-1}{2}x^{\frac{-3}{2}}$. Note that you always subtract 1 from the exponent, even when it is negative. Written in exponent form, x had an exponent of $\dfrac{-1}{2}$, and $\dfrac{-1}{2} - 1 = \dfrac{-3}{2}$. Compare the work done here with that of Example 4.6. Notice that the answer given there, $\dfrac{-1}{2x\sqrt{x}}$, is the same as that found here because $x\sqrt{x} = x \cdot x^{\frac{1}{2}} = x^{\frac{3}{2}}$, since exponents add.

Example 5.7 Find the derivative of $g(t) = \dfrac{1}{t^2}$.

> *Solution:* As $g(t) = \dfrac{1}{t^2} = t^{-2}$, the power rule gives $g'(t) = -2t^{-3} = \dfrac{-2}{t^3}$. Except for the change in the letters, this is the same answer as obtained in Example 4.7.

Remember that to use the power rule effectively, you must sometimes convert $\sqrt{}$ signs to exponents of $\dfrac{1}{2}$, and fractional expressions such as $\dfrac{1}{x^{\frac{1}{2}}}$ to expressions with a negative exponent such as $x^{\frac{-1}{2}}$.

5.2 DERIVATIVES OF COMBINATIONS OF FUNCTIONS

Suppose that you have two functions $f(x)$ and $g(x)$ and that you have already computed their derivatives $f'(x)$ and $g'(x)$. In this section you will find the formulas, in terms of f, f', g, and g', for the derivatives of functions formed by arithmetic combinations of f and g, such as $f + g$, $\dfrac{f}{g}$, and so on. Again, the section will go lightly over proofs, but will emphasize learning the formulas in terms of *words*.

We make the blanket assumption in this section that the functions f and g are differentiable at x. It will also be an implicit conclusion of each theorem that the compound function is differentiable at x. The only exception to this will be in the formula for the derivative of $\dfrac{f}{g}$, where the problem of division by 0 might arise.

THEOREM 5.3. *Let c be a constant and $f(x)$ be a function.*

1. *Let the function F be defined by $F(x) = cf(x)$. Then*

$$F'(x) = cf'(x)$$

More concisely,

$$(cf)' = cf'$$

2. *If $u = f(x)$, then $\dfrac{d(cu)}{dx} = c \cdot \dfrac{du}{dx}$.*

3. *The derivative of a constant times a function equals the constant times the derivative of the function. In other words, when computing derivatives, multiplicative constants can be pulled out of the expression.*

Example 5.8 Find the derivative of $s(t) = 7t^3$.

Solution:

$$\begin{aligned} (7t^3)' \ &= 7 \cdot (t^3)' \quad \text{(Theorem 5.3)} \\ &= 7 \cdot 3t^2 \quad \text{(the power rule)} \\ &= 21t^2 \end{aligned}$$

THEOREM 5.4. *Let $f(x)$ and $g(x)$ be functions.*

1. *Let the function F be defined by $F(x) = f(x) + g(x)$. Then*

$$F'(x) = f'(x) + g'(x)$$

More concisely,

$$(f + g)' = f' + g'$$

2. *If $u = f(x)$ and $v = g(x)$, then $\dfrac{d(u + v)}{dx} = \dfrac{du}{dx} + \dfrac{dv}{dx}$.*

3. *The derivative of a sum is the sum of the derivatives.*

Example 5.9 Find the derivative of $f(x) = x^2 + x^3$.

Solution:

$$\begin{aligned} \frac{d(x^2 + x^3)}{dx} \ &= \frac{d(x^2)}{dx} + \frac{d(x^3)}{dx} \qquad \text{(Theorem 5.4)} \\ &= 2x + 3x^2 \qquad \text{(the power rule)} \end{aligned}$$

Example 5.10 Find the derivative of $g(t) = 7t^3 + 6t^2$.

> *Solution:* You just use the different rules as the need for them arises. First of all, g is a sum of two terms, so you use Theorem 5.4 first, and then other rules as appropriate on the individual summands.

$$
\begin{aligned}
(7t^3 + 6t^2)' &= (7t^3)' + (6t^2)' && \text{(Theorem 5.4)} \\
&= 7 \cdot (t^3)' + 6 \cdot (t^2)' && \text{(Theorem 5.3)} \\
&= 7 \cdot 3t^2 + 6 \cdot 2t && \text{(the power rule)} \\
&= 21t^2 + 12t
\end{aligned}
$$

You can probably see that with a little practice, you would be able to write the derivative in just one step, using all the rules at once. Be patient in this section and try to use only one rule at a time. That will help you learn the rules, and also force you to think about how more complicated formulas are built up from their simpler pieces. After the practice of this section, feel free to combine as many steps as you can.

THEOREM 5.5. *Let $f(x)$ and $g(x)$ be functions.*

1. *Let the function F be defined by $F(x) = f(x) - g(x)$. Then*

$$F'(x) = f'(x) - g'(x)$$

> *More concisely,*

$$(f - g)' = f' - g'$$

2. *If $u = f(x)$ and $v = g(x)$, then* $\dfrac{d(u - v)}{dx} = \dfrac{du}{dx} - \dfrac{dv}{dx}$

3. *The derivative of a difference is the difference of the derivatives.*

Example 5.11 Find the derivative of $x^2 - 6x$.

> *Solution:*

$$
\begin{aligned}
\frac{d}{dx}(x^2 - 6x) &= \frac{d}{dx}(x^2) - \frac{d}{dx}(6x) && \text{(Theorem 5.5)} \\
&= 2x - 6\frac{dx}{dx} && \text{(the power rule and Theorem 5.3)} \\
&= 2x - 6 \cdot 1 = 2x - 6
\end{aligned}
$$

Compare the above work with that needed to solve Example 4.3 of Section 4.1.

With the three preceding theorems, you can find the derivative of any polynomial.

Example 5.12 Find the derivative of $p(t) = 11t^4 - 6t^3 + 9t^2 + 23t - 5$.

Solution: The derivatives of sums and/or differences are the sums and/or differences of the derivatives. Thus,

$$
\begin{aligned}
(11t^4 - 6t^3 + 9t^2 + 23t - 5)' &= (11t^4)' - (6t^3)' + (9t^2)' + (23t)' - (5)' \\
&= 11 \cdot (t^4)' - 6 \cdot (t^3)' + 9 \cdot (t^2)' + 23 \cdot (t)' - 0 \\
&= 11 \cdot 4t^3 - 6 \cdot 3t^2 + 9 \cdot 2t + 23 \cdot 1 \\
&= 44t^3 - 18t^2 + 18t + 23
\end{aligned}
$$

You can also fairly easily find the derivatives of certain fractional expressions, if you are willing to do a little algebra.

Example 5.13 Find the derivative of $\dfrac{x^3 + 1}{x^2}$.

Solution: By doing some preliminary algebra, you can rewrite the function as a sum of powers:

$$
\begin{aligned}
\frac{x^3 + 1}{x^2} &= \frac{x^3}{x^2} + \frac{1}{x^2} \\
&= x^{(3-2)} + x^{-2} \\
&= x^1 + x^{-2} \\
&= x + x^{-2}
\end{aligned}
$$

Thus,

$$
\begin{aligned}
\left(\frac{x^3 + 1}{x^2}\right)' &= \left(x + x^{-2}\right)' \\
&= 1 + (-2x^{-3}) \\
&= 1 - \frac{2}{x^3}
\end{aligned}
$$

The proofs of each of the preceding three theorems follow in a fairly straightforward manner from the definition of derivative and the fact discussed in Chapter 3, "The Idea of Limits," that limits behave reasonably with respect to the operations of arithmetic. (See Section 6 of Chapter 3, "Summary of Main Points.") The following is a typical proof. Recall that in Theorem 5.4, $F(x) = f(x) + g(x)$.

Proof of Theorem 5.4:

$$
\begin{aligned}
F'(x) &= \lim_{h \to 0} \frac{F(x+h) - F(x)}{h} & \text{(definition of derivative)}\\[2mm]
&= \lim_{h \to 0} \frac{(f(x+h) + g(x+h)) - (f(x) + g(x))}{h} & \text{(definition of } F(x))\\[2mm]
&= \lim_{h \to 0} \frac{(f(x+h) - f(x)) + (g(x+h) - g(x))}{h} & \text{(commutativity of addition)}\\[2mm]
&= \lim_{h \to 0} \left(\frac{f(x+h) - f(x)}{h} + \frac{g(x+h) - g(x)}{h} \right)\\[2mm]
&= \lim_{h \to 0} \frac{f(x+h) - f(x)}{h} + \lim_{h \to 0} \frac{g(x+h) - g(x)}{h} & \text{(limits behave reasonably with respect to addition)}\\[2mm]
&= f'(x) + g'(x) & \text{(definition of derivative)}
\end{aligned}
$$

The next two theorems describe the formulas for derivatives of products and quotients. They are more complicated than those for sums and differences.

THEOREM 5.6. *(The Product Rule) Let* $f(x)$ *and* $g(x)$ *be functions.*

1. *Let the function F be defined by* $F(x) = f(x)g(x)$. *Then*

$$
F'(x) = f(x)g'(x) + f'(x)g(x)
$$

   More concisely,

$$
(fg)' = fg' + f'g
$$

2. *Let* $u = f(x)$ *and* $v = g(x)$. *Then* $\dfrac{d(uv)}{dx} = u\dfrac{dv}{dx} + \dfrac{du}{dx}v$

3. *The derivative of a product equals the first factor times the derivative of the second, plus the derivative of the first factor, times the second.*

I will explain shortly why the product rule holds, but first let's see an example of its use.

Example 5.14 Find the derivative of $f(x) = (x^2 - x^3)(x^4 + 2x)$ without first multiplying out the factors.

Solution: The first factor is $x^2 - x^3$, with derivative $2x - 3x^2$, and the second factor is $x^4 + 2x$, with derivative of $4x^3 + 2$. Thus the first factor times the derivative of the second, plus the derivative of the first factor, times the second, equals

$$
(x^2 - x^3)(4x^3 + 2) + (2x - 3x^2)(x^4 + 2x)
$$

The product rule at first glance seems surprising enough that you should see some idea of why it is true. Recall that in Theorem 5.6, $F(x) = f(x)g(x)$.

Proof of Theorem 5.6:

$$F'(x) = \lim_{h \to 0} \frac{F(x+h) - F(x)}{h} \qquad \text{(definition of derivative)}$$

$$= \lim_{h \to 0} \frac{f(x+h)g(x+h) - f(x)g(x)}{h} \qquad \text{(definition of } F(x)) \quad (5.1)$$

Now unlike the proof of the formula for the derivative of sums, there is no obvious way to rearrange the above numerator to separate the f terms from the g terms. Rearranging by factoring doesn't seem possible because none of the four different terms in the numerator are the same. The technique that allows you to proceed is to add to the numerator, and then subtract from the numerator, the same expression (so the numerator remains unchanged in value). This expression is chosen to be a product, with one f term and one g term, and with one of the terms being evaluated at $x + h$ like the first pair of terms in the numerator, and the other being evaluated at x like the second pair of terms in the numerator. Such a choice allows you to rearrange the numerator by factoring, and to separate out the f terms from the g terms. Subtract and add $f(x+h)g(x)$ to the numerator. Equation 5.1 becomes

$$\lim_{h \to 0} \frac{f(x+h)g(x+h) - f(x+h)g(x) + f(x+h)g(x) - f(x)g(x)}{h}$$

$$= \lim_{h \to 0} \left(f(x+h)\frac{g(x+h) - g(x)}{h} + g(x)\frac{f(x+h) - f(x)}{h} \right)$$

$$= \lim_{h \to 0} f(x+h) \lim_{h \to 0} \frac{g(x+h) - g(x)}{h} + g(x) \lim_{h \to 0} \frac{f(x+h) - f(x)}{h}$$

$$= f(x)g'(x) + g(x)f'(x)$$

In going from the next to the last line to the last line, $\lim_{h \to 0} f(x + h) = f(x)$ because f is assumed to be differentiable at x, and differentiable functions are continuous by Theorem 3.1. Therefore, as $h \to 0$, $x + h \to x$ and by continuity $f(x + h) \to f(x)$. The other two limits just involve the definition of the derivative.

The following theorem gives the rule for the derivative of a quotient of two functions. The statement in words sounds too stiff if you use "numerator" and "denominator," so replace these terms simply with "top" and "bottom."

THEOREM 5.7. *(The Quotient Rule) Let $f(x)$ and $g(x)$ be functions, with $g(x) \neq 0$.*

1. *Let the function F be defined by* $F(x) = \dfrac{f(x)}{g(x)}$. *Then*

$$F'(x) = \frac{g(x)f'(x) - f(x)g'(x)}{[g(x)]^2}$$

More concisely,

$$\left(\frac{f}{g}\right)' = \frac{gf' - fg'}{g^2}$$

2. *If* $u = f(x)$ *and* $v = g(x)$, *then*

$$\frac{d\left(\frac{u}{v}\right)}{dx} = \frac{v\frac{du}{dx} - u\frac{dv}{dx}}{v^2}$$

3. *The derivative of a quotient is the bottom times the derivative of the top, minus the top times the derivative of the bottom, all over the bottom squared.*

The following example is the same as Example 5.13, but instead of first simplifying the function algebraically, it uses the quotient rule. It's left up to you to do the algebra needed to check that both answers are really the same.

Example 5.15 Find the derivative of $\dfrac{x^3 + 1}{x^2}$.

> **Solution:** The top $x^3 + 1$ has derivative $3x^2$, and the bottom x^2 has derivative $2x$, so the bottom times the derivative of the top, minus the top times the derivative of the bottom, all over the bottom squared, equals
>
> $$\frac{x^2 \cdot 3x^2 - (x^3 + 1) \cdot 2x}{(x^2)^2}$$

The quotient rule lets you *prove* the power rule for negative integers. Here is an example:

$$
\begin{aligned}
\left(x^{-4}\right)' &= \left(\frac{1}{x^4}\right)' \\
&= \frac{x^4 \cdot 1' - 1 \cdot \left(x^4\right)'}{\left(x^4\right)^2} \\
&= \frac{x^4 \cdot 0 - 1 \cdot 4x^3}{x^8} \\
&= \frac{0 - 4x^3}{x^8} \\
&= \frac{-4}{x^5} \\
&= -4x^{-5}
\end{aligned}
$$

This is exactly what the power rule says it should be.

The quotient rule can be proven in two steps. First, as a special case, you can get the formula for $\left(\dfrac{1}{g(x)}\right)'$ in terms of g and g'. This can be done much as you computed the derivatives of fractional expressions with the variable in the denominator, in Section 4.2, by using common denominators. Secondly, you can use the product rule on $\dfrac{f(x)}{g(x)} = f(x) \cdot \dfrac{1}{g(x)}$.

Example 5.16 Find the derivative of $f(x) = \dfrac{(x^2 + 1)(x^3 + 1)}{2x}$.

> **Solution:** Of course, you could multiply out the terms in the numerator and then just use the quotient rule, but try to use just the derivative rules on the function as written. It is first of all a quotient, in which the numerator is a product. So first of all use the quotient rule, and when you must find the derivative of the numerator, use the product rule.

$$
\begin{aligned}
f'(x) &= \frac{2x[(x^2 + 1)(x^3 + 1)]' - [(x^2 + 1)(x^3 + 1)](2x)'}{(2x)^2} \\[2mm]
&= \frac{2x[(x^2 + 1)(x^3 + 1)' + (x^2 + 1)'(x^3 + 1)] - [(x^2 + 1)(x^3 + 1)](2x)'}{(2x)^2} \\[2mm]
&= \frac{2x[(x^2 + 1)3x^2 + 2x(x^3 + 1)] - [(x^2 + 1)(x^3 + 1)]2}{4x^2}
\end{aligned}
$$

5.3 THE CHAIN RULE

There is one more standard way of combining two functions f and g, namely, **composition**. To understand composition of functions, it is extremely useful to think of a function as giving the procedure, or recipe, for taking a number and computing an associated value. For example, if $f(x) = x^2 + 1$, the procedure is to take a number, square it, and then add 1. Think in terms of punching buttons on a calculator. If you were asked to compute $f(3.5)$, you would enter 3.5, punch the squaring button to get 12.25, and then add 1 to get 13.25, so that $f(3.5) = 13.25$. When two functions are composed, the value of one of the functions, at some point, becomes the number on which the second function performs its procedure. Thus, if $g(x) = \sqrt{x}$, and you were asked to compute the value of g *composed with* f at $x = 3.5$, you would first compute $f(3.5) = 13.25$ as above, and then with 13.25 showing in the calculator window, punch the square root button.

DEFINITION 5.1. *Let $f(x)$ and $g(x)$ be functions. The* **composition of g and f**, *denoted by $g \circ f$, is defined by $(g \circ f)(x) = g(f(x))$.*

The expression $(g \circ f)(x)$ is read as "g composed with f, at x," while $g(f(x))$ is read as "g of f at x" or "g of f of x."

Example 5.17 Let $f(x) = x^2 + 1$ and $g(x) = \sqrt{x}$, as above. Find the formulas for $(g \circ f)(x)$ and $(f \circ g)(x)$.

> **Solution:** $(g \circ f)(x) = g(f(x)) = g(x^2 + 1) = \sqrt{x^2 + 1}$, while $(f \circ g)(x) = f(\sqrt{x}) = (\sqrt{x})^2 + 1 = x + 1$. This example shows that $(g \circ f)(x)$ need not be equal to $(f \circ g)(x)$.

Remember the old rule about the order in which to perform operations—do what is in the parentheses first. In $g(f(x))$, the $f(x)$ is in parentheses and the function g is applied to $f(x)$. In evaluating $g(f(x))$ at a particular value, such as $x = 3.5$, on a standard calculator, you would first evaluate $f(3.5)$ and then apply g to $f(3.5)$. Often, in the formula for $(g \circ f)(x) = g(f(x))$, the $f(x)$ actually appears inside parentheses. In $g(f(x))$, the $f(x)$, the part to be computed first, will be called the *inside* part, and the g, the part to be computed second, will be called the *outside* part.

In order for you to be able to use the chain rule, the rule for the derivative of a composition of two functions, effectively, you must be able to look at a formula and see how it is composed of two simpler functions. One way of doing this is by asking yourself how you would evaluate the function at a particular value of x on a calculator. The steps you go through first would describe what the inside function should be, and the steps you go through last would describe what the outside function should be. If your formula has parentheses, then probably the expression inside the parentheses would be the inside part, and the rest would be the outside.

Example 5.18 Write the function $h(x) = (1 + x^2)^3$ as the composition $g \circ f$ of two simpler functions.

> **Solution:** Clearly the part inside the parentheses is $1 + x^2$, and that is what you would calculate first if you had to compute h at a particular value. You would then apply the cubing function to, or cube, the value of $1 + x^2$ to get $h(x)$. Thus the inside part would be $f(x) = 1 + x^2$, and the outside part would be $g(x) = x^3$, and $h(x) = (g \circ f)(x) = g(f(x)) = g(1 + x^2) = (1 + x^2)^3$.

Example 5.19 Write the function $s(t) = \sqrt{3t^2 + 4}$ as the composite $g \circ f$ of two simpler functions.

> **Solution:** There are no parentheses in the formula for s, but there are natural inside and outside parts. If you were asked to compute $s(7)$ on a calculator, you would first compute $3 \cdot 7^2 + 4 = 151$, and then press the square root button. What you compute first is the inside part, and what you compute next is the outside part, so with $f(t) = 3t^2 + 4$ and $g(t) = \sqrt{t}$, $h(t) = g(f(t)) = (g \circ f)(t)$.

Let $y = g(f(x))$. The term $f(x)$ plays two roles—it is of course a function, but it is sort of like a variable, in that it provides the input to the function g. In dealing with the derivatives of composite functions, a common technique is to highlight these two roles by introducing a new letter u. Let $u = f(x)$, so that $y = g(f(x)) = g(u)$. Think about the equations

$$y = g(u) \text{ and } u = f(x)$$

You have that y depends upon u, while u depends upon x. If you think about chaining or composing these two dependencies together, then of course y also depends upon x. The symbol $\dfrac{dy}{dx}$ means that you are viewing y as a function of x, and taking the derivative of the function y with respect to the variable x. The symbol $\dfrac{dy}{du}$ means that you are viewing y as a function of u, and taking the derivative of the function y with respect to the variable u.

Example 5.20 Let $y = (1 + x^2)^3$. Write y as a simpler function of a new variable u, where u is a function of x.

Solution: Let $u = 1 + x^2$. Then in terms of u, $y = u^3$. Compare this to Example 5.18. Clearly a good choice for u is just the inside part of the composite.

Example 5.21 Let $y = \sqrt{u}$, where $u = 3t^2 + 4$. Write the equation for y as a function of t.

Solution: Simply substitute $u = 3t^2 + 4$ into $y = \sqrt{u}$ to get $y = \sqrt{3t^2 + 4}$.

THEOREM 5.8. *(The Chain Rule) Let $f(x)$ and $g(x)$ be functions, with f differentiable at x, and g differentiable at the point $f(x)$.*

1. *Let the function F be defined by $F = g \circ f$, so that $F(x) = g(f(x))$. Then F is differentiable at x, and*

$$F'(x) = g'(f(x)) \cdot f'(x)$$

2. *Let $y = g(u)$ and let $u = f(x)$. Of course, y can also be considered as a function of x, since $y = g(u) = g(f(x))$.*

$$\frac{dy}{dx} = \frac{dy}{du} \cdot \frac{du}{dx}$$

3. *The derivative of a composite function equals the derivative of the outside function, evaluated at the inside part, times the derivative of the inside part.*

More than the other theorems of this chapter, the chain rule requires an explanation on how to apply the formula. In the following example, the function of Examples 5.18 and 5.20 will be used to illustrate each of the above three ways of looking at the chain rule. Note that although some of the intermediate work and symbols might look different in each of the three methods, the final answer is the same (as it should be).

Example 5.22 Use the chain rule to find the derivative of $h(x) = (1 + x^2)^3$.

Solution:

1. As in Example 5.18, you write h as the composite of two simpler functions g and f, so that $h(x) = (g \circ f)(x) = g(f(x))$. The two simpler functions here are $f(x) = 1 + x^2$ and $g(x) = x^3$. Now the chain rule says that

$$h'(x) = g'(f(x)) \cdot f'(x)$$

What is $g'(f(x))$? Well, as $g(x) = x^3, g'(x) = 3x^2$, by the power rule. Thus,

$$g'(f(x)) = g'(1 + x^2) = 3(1 + x^2)^2$$

As $f'(x) = 2x$, you have that

$$g'(f(x)) \cdot f'(x) = 3(1 + x^2)^2 \cdot 2x = 6x(1 + x^2)^2$$

2. As in Example 5.20, introduce $u = 1 + x^2$, so that

$$y = u^3 \text{ and } u = 1 + x^2$$

The second formulation of the chain rule says that

$$\frac{dy}{dx} = \frac{dy}{du} \cdot \frac{du}{dx}$$

Now y is simply a power of u, so that $\dfrac{dy}{du}$ can be easily computed by the power rule, to give $\dfrac{dy}{du} = 3u^2$.

$$
\begin{aligned}
\frac{dy}{dx} &= \frac{dy}{du} \cdot \frac{du}{dx} \\
&= \frac{d(u^3)}{du} \cdot \frac{d(1 + x^2)}{dx} \\
&= 3u^2 \cdot 2x \\
&= 3(1 + x^2)^2 \cdot 2x
\end{aligned}
$$

Note to students:

  (a) The purpose of the final line above is to get the derivative solely in terms of the original variable x.

  (b) Also, compare the work in the above two methods to see that they are not as dissimilar as they might first appear.

  (c) Although derivatives are not fractions, you can be sure you have this version of the chain rule written down correctly if it just looks correct— in other words, if you *thought of* the derivatives as fractions, then in looking at $\dfrac{dy}{du} \cdot \dfrac{du}{dx}$, it *looks* as though du would cancel from the numerator and denominator to give $\dfrac{dy}{dx}$.

3. With a little practice and experience, you should be able to use the ditty "the derivative of the outside function, evaluated at the inside part, times the derivative of the inside part" to compute the derivatives of composite functions. You should not have to explicitly write out the simpler functions f and g, or the intermediate variable u. To illustrate the method, look at $h = (1 + x^2)^3$. The first derivative to compute is the "derivative of the outside function." In doing this, it does not matter what the inside function is, and you can just ignore the inside function. The outside function is

$$(\ )^3$$

The pattern for the derivative of this function is, by the power rule:

$$3(\ )^2$$

Now the derivative of the outside function, evaluated at the inside part, is

$$3(1 + x^2)^2$$

Thus, the final answer, the derivative of the outside function, evaluated at the inside part, times the derivative of the inside part, is

$$3(1 + x^2)^2 \cdot 2x$$

I will work out one more example by all three methods (but with shorter explanations) and then several more examples by using the ditty.

Example 5.23 Find the derivative of $s(t) = \sqrt{3t^2 + 4}$.

Solution: First, to take derivatives, replace the $\sqrt{}$ sign with a fractional exponent.

1. You can write $s(t) = g(f(t))$, where $f(t) = 3t^2 + 4$ and $g(t) = \sqrt{t} = t^{\frac{1}{2}}$. As $f'(t) = 6t$ and $g'(t) = \frac{1}{2}t^{-\frac{1}{2}}$, it follows that

$$
\begin{aligned}
s'(t) &= g'(f(t)) \cdot f'(t) \\
&= \frac{1}{2}(f(t))^{-\frac{1}{2}} \cdot 6t \\
&= \frac{1}{2}(3t^2 + 4)^{-\frac{1}{2}} \cdot 6t \\
&= 3t(3t^2 + 4)^{-\frac{1}{2}}
\end{aligned}
$$

2. Let $y = u^{\frac{1}{2}}$, with $u = 3t^2 + 4$. Then $\dfrac{dy}{du} = \dfrac{1}{2}u^{-\frac{1}{2}}$ and $\frac{du}{dt} = 6t$. It follows that

$$
\begin{aligned}
\frac{dy}{dt} &= \frac{dy}{du} \cdot \frac{du}{dt} \\
&= \frac{1}{2}u^{-\frac{1}{2}} \cdot 6t \\
&= \frac{1}{2}(3t^2 + 4)^{-\frac{1}{2}} \cdot 6t \\
&= 3t(3t^2 + 4)^{-\frac{1}{2}}
\end{aligned}
$$

3. At first, ignore the inside part, so that $h = (\ )^{\frac{1}{2}}$. The pattern for the derivative of the outside function gives $\frac{1}{2}(\ )^{-\frac{1}{2}}$, and the derivative of the outside function, evaluted at the inside part, is $\frac{1}{2}(3t^2 + 4)^{-\frac{1}{2}}$. Finally, the derivative of the outside function, evaluated at the inside part, times the derivative of the inside part, gives

$$
\frac{1}{2}(3t^2 + 4)^{-\frac{1}{2}} \cdot 6t = 3t(3t^2 + 4)^{-\frac{1}{2}}
$$

Example 5.24 Find the derivative of $f(t) = \sqrt[3]{2t}$.

Solution: First of all, change the symbol $\sqrt[3]{}$ for the cube root to exponent form, so that $f(t) = (2t)^{\frac{1}{3}}$. The outside function $(\ )^{\frac{1}{3}}$

has derivative with the pattern $\frac{1}{3}(\ )^{-\frac{2}{3}}$ by the power rule, so the derivative equals

$$\frac{1}{3}(2t)^{-\frac{2}{3}} \cdot 2$$

Do not forget to multiply by the derivative of the inside part, which is where the final 2 in the preceding example came from. When the inside part is rather simple, as it was here, it is easier to forget the final step of the chain rule—"times the derivative of the inside part."

Example 5.25 Find the derivative of $f(x) = \left(\dfrac{1+4x}{1-4x}\right)^3$.

> *Solution:* The function is a composite, with the outside part being the cubing function, and the inside part a quotient. You will have to use the quotient rule when computing the derivative of the inside part, but you don't have to worry about that when starting the chain rule—first focus on the outside part. Now $f'(x) = 3(\ )^2 \cdot$ $(\ )'$. The final $\cdot\ (\ )'$ is to remind yourself to multiply by the derivative of the inside part. Now fill in the inside part—$f'(x) =$ $3\left(\dfrac{1+4x}{1-4x}\right)^2 \cdot (\ )'$. Now compute the derivative of the inside part by the quotient rule, to get
>
> $$\begin{aligned} f'(x) &= 3\left(\frac{1+4x}{1-4x}\right)^2 \cdot \left(\frac{(1-4x)4 - (1+4x)(-4)}{(1-4x)^2}\right) \\ &= 3\left(\frac{1+4x}{1-4x}\right)^2 \cdot \left(\frac{8}{(1-4x)^2}\right) \\ &= \frac{24(1+4x)^2}{(1-4x)^4} \end{aligned}$$

Example 5.26 Find the derivative of $f(x) = \dfrac{7x - (x^2+1)^5}{2x-7}$.

> *Solution:* To use the derivative formulas correctly, you have to see how a formula is made up from its simpler pieces. The above, as written, is first of all a quotient, and you must use the quotient rule. When finding the derivative of the numerator, you must see that the numerator is first of all a difference, and use the fact that the derivative of a difference equals the difference of the

derivatives. Only when finding the derivative of the $(x^2+1)^5$ part of the numerator would you use the chain rule. Thus,

$$f'(x) = \frac{(2x-7)[7x-(x^2+1)^5]' - [7x-(x^2+1)^5](2x-7)'}{(2x-7)^2}$$

$$= \frac{(2x-7)[(7x)' - ((x^2+1)^5)'] - [7x-(x^2+1)^5]((2x)'-7')}{(2x-7)^2}$$

$$= \frac{(2x-7)[7 - 5(x^2+1)^4 \cdot (x^2+1)'] - [7x-(x^2+1)^5]2}{(2x-7)^2}$$

$$= \frac{(2x-7)[7 - 5(x^2+1)^4 \cdot 2x] - [7x-(x^2+1)^5]2}{(2x-7)^2}$$

$$= \frac{(2x-7)[7 - 10x(x^2+1)^4] - [7x-(x^2+1)^5]2}{(2x-7)^2}$$

Example 5.27 Find the derivative of $f(x) = (1 + \sqrt{x^2+1})^3$.

Solution: First of all, change the square root sign to a fractional exponent, so that $f(x) = [1 + (x^2+1)^{\frac{1}{2}}]^3$. Clearly the outside function is the cubing function, and the inside function is $1 + (x^2+1)^{\frac{1}{2}}$. To compute the derivative of the inside function, you must use the chain rule again, but don't worry about that now. The outside function has the pattern of $(\ )^3$, and the derivative thus has the pattern of $3(\ )^2$, so the derivative of the outside function, evaluated at the inside part, is $3[1 + (x^2+1)^{\frac{1}{2}}]^2$. To compute the derivative of the inside part $1 + (x^2+1)^{\frac{1}{2}}$ you must use the rule for the derivative of sums, as well as the chain rule again. In computing the derivative of $(x^2+1)^{\frac{1}{2}}$, the function with the pattern $(\ )^{\frac{1}{2}}$ is now the outside part, and the function x^2+1 is the inside part. Thus,

$$[1 + (x^2+1)^{\frac{1}{2}}]' = 0 + \tfrac{1}{2}(x^2+1)^{-\frac{1}{2}} \cdot 2x \quad \text{(the } 2x \text{ comes from the}$$
final derivative of the inside part)

$$= x(x^2+1)^{-\frac{1}{2}}$$

$$= \frac{x}{\sqrt{x^2+1}}$$

Finally,

$$f'(x) = 3\left[1 + (x^2 + 1)^{\frac{1}{2}}\right]^2 \cdot \frac{x}{\sqrt{x^2 + 1}} = \frac{3x\left[1 + \sqrt{x^2 + 1}\right]^2}{\sqrt{x^2 + 1}}$$

5.4 SUMMARY OF MAIN POINTS

- The derivative of a constant equals 0.

- (The Power Rule) To find the derivative of a power of the variable, simply multiply in front by the exponent and subtract 1 from the exponent.

- To make derivatives easier to compute by means of the power rule, always change $\sqrt{}$ in a formula to the $\frac{1}{2}$ power. If it is not too complicated, change easy fractional expressions with the variable in the denominator to negative exponent form, such as changing $\frac{1}{x^3}$ to x^{-3}.

- The derivative of a constant times a function equals the constant times the derivative of the function. In computing derivatives, you can think of multiplicative constants as sliding out past the derivative sign. For example, $(7x^2)' = 7 \cdot (x^2)'$.

- The derivative of a sum is the sum of the derivatives.

- The derivative of a difference is the difference of the derivatives.

- (The Product Rule) The derivative of a product equals the first factor times the derivative of the second, plus the derivative of the first factor, times the second.

- (The Quotient Rule) The derivative of a quotient equals the bottom times the derivative of the top, minus the top times the derivative of the bottom, all over the bottom squared.

- (The Chain Rule) The derivative of a composite function equals the derivative of the outside function, evaluated at the inside part, all times the derivative of the inside part.

- To use the derivative formulas correctly, you must be able to see how a formula is put together from its simpler pieces.

5.5 EXERCISES

Find the derivatives of the following functions.

1. $f(x) = 2x^2 - 3x + 5$

2. $g(t) = 7t^3 - 4t^2 + \dfrac{1}{2}t - 9$

3. $p(s) = 6s^3 - 3\sqrt{s} + \dfrac{4}{s} - \dfrac{11}{s^2}$

4. $h(u) = -4u^{79}$

5. $g(x) = \dfrac{x^3 + x + 1}{x^2}$

6. $f(x) = 3x^{\frac{2}{3}} + 4\sqrt{x}$

7. $s(t) = \dfrac{7}{t^3} + \dfrac{1}{3t^2}$

8. $p(u) = \sqrt{u} - \dfrac{1}{\sqrt{u}}$

9. $f(x) = (7x^5 - 5x^3 + 6x^2 - 4) \cdot (8x^4 - 9x^2 + 24)$

10. $f(t) = \sqrt{t} \cdot (t^2 - t)$

11. $h(x) = \dfrac{x^2 + x}{x^2 - x + 1}$

12. $p(t) = \dfrac{3t^3 + t}{t^2 - 1}$

13. $f(x) = \dfrac{2x^2 - 4x + 5}{3\sqrt{x} + 1}$

14. $h(v) = \dfrac{3}{\sqrt{5v - 2}}$

15. $f(s) = s^2 \cdot (3s^2 - 6s + 9)^4$

16. $p(t) = \left(\dfrac{t^2 - 3t}{2t^2 + 5t} \right)^3$

17. $f(x) = \sqrt{2x}$

18. $h(x) = \sqrt{7x^3 + 4x^2}$

19. $p(t) = (2t)^{34}$

20. $g(t) = \dfrac{(t^2 - 4t) \cdot (4t + 1)}{3t^2 - 3}$

21. $h(s) = \left(\sqrt{5s^2 + 1} + s^3 \right)^4$

22. $f(x) = x^{\frac{4}{3}} - \dfrac{1}{x^{\frac{4}{3}}}$

23. $p(t) = (3t^2 - 1)^3 \cdot (4t + 9)^5$

24. $h(x) = \dfrac{(x^2 + 1)^4}{(2x - 3)^7}$

25. $f(x) = [3x - (x^4 + x)^5]^3$

6

EXTREME VALUES, THE MEAN VALUE THEOREM, AND CURVE SKETCHING

Many important and practical real-world applications of calculus involve figuring out how to perform certain procedures as efficiently as possible. Examples might involve finding the smallest possible cost or the shortest possible time for performing a given procedure or task, or figuring out how to perform a task most productively under a given budget and time schedule. The basic mathematical question underlying such applied problems is how to find the largest or smallest values of a given function on a given interval. This chapter tackles the underlying mathematical question, while Section 7.2 deals with the applied problems themselves.

The procedures for finding the largest and smallest values of a function f on an interval, and even whether the function has largest and smallest values on that interval, depend upon the nature of the interval. Section 6.1 deals primarily with continuous functions on an interval that is both **closed** and **bounded**. Such intervals have finite length and contain both of their endpoints, and are written, in interval notation, with brackets rather than parentheses. An example is the interval $[2, 5]$. It has finite length of $5 - 2 = 3$, and the bracket notation means that the endpoints 2 and 5 are considered to be part of the interval. Thus, a number x lies in the interval $[2, 5]$ if and only if x satisfies the inequality $2 \leq x \leq 5$. Closed and bounded intervals are distinguished from other types of intervals, which may have finite length, but not contain their endpoints, such as the interval $(2, 5)$, which consists of all x satisfying the inequality $2 < x < 5$, and intervals that have infinite length, such as $(0, \infty)$, consisting of all x satisfying the equation $x > 0$.

In studying the question of largest and smallest values of a continuous function f on an interval that is *not* closed and bounded, it is usually the case that more attention must be paid to the behavior of the graph of f, and specifically to where the graph is rising and where it is falling. Calculus also helps with this, in a manner that is intuitively pretty obvious, but that actually requires, for justification, a very important and powerful theorem, called the Mean Value Theorem. Section 6.2 deals with this theorem and its applications. Section 6.3 deals more specifically with the techniques of how to use the results of Section 6.2 to sketch the graph of a function.

6.1 EXTREME VALUES

DEFINITION 6.1. *The largest value a function has on an interval is called its* **maximum value**. *The smallest value a function has on an interval is called its* **minimum value**. *Both the maximum value and the minimum value are called* **extreme values**.

A basic fact is that a continuous function *on a closed and bounded interval* $[a, b]$ does always have a maximum and minimum value. A rigorous proof is beyond the scope of a standard first-year calculus course. Intuitively, though, it is because you have to start drawing the graph at the point $(a, f(a))$ and end by drawing it at the point $(b, f(b))$, without lifting your pencil off the paper. It is certainly hard to imagine how you could do this without having some point on the graph be a high point (maximum value), and some other point be a low point (minimum value). (Note that on a horizontal line, all points would be considered to be simultaneously high points and low points.) An example is given by the graph of $f(x) = x^2$ on the interval $[-1, 2]$, pictured in Figure 6.1(a). The filled black circles are there to emphasize the fact that the endpoints are included in the interval and thus on the graph. Clearly the minimum value (low point) occurs at the point $x = 0$, and the minimum value is $f(0) = 0^2 = 0$. The maximum value (high point) occurs at the right-hand endpoint $x = 2$, and the maximum value is $f(2) = 2^2 = 4$.

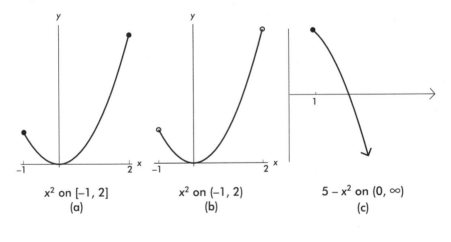

x^2 on $[-1, 2]$ x^2 on $(-1, 2)$ $5 - x^2$ on $(0, \infty)$
(a)   (b)   (c)

Figure 6.1

Figures 6.1(b) and 6.1(c) provide examples of how a continuous function may not have maximum and/or minimum values on an interval that is not closed and bounded. In (b) is the graph of $f(x) = x^2$ on the interval $(-1, 2)$. The hollow circles are there to emphasize the fact that the endpoints are *not* included, either in the interval or on the graph. Clearly the graph still has a low point, which occurs at $x = 0$. Thus the function $f(x) = x^2$ on the interval $(-1, 2)$ does have a minimum value of $f(0) = 0^2 = 0$. However, the graph has no high point. Remember that the point $(2, 4)$ is *not* on the graph. For any point you pick on the graph, there is another point that is higher. Thus, $f(x) = x^2$ does *not* have a maximum value on the interval $(-1, 2)$. The problem is sort of that the graph is not pinned down at the endpoint $x = 2$.

A further example is provided by the graph of $f(x) = 5 - x^2$ on the interval $[1, \infty)$, pictured in Figure 6.1(c). This graph certainly has a high point, which occurs at $x = 1$. Thus the function has a maximum value on the interval $[1, \infty]$ of $f(1) = 5 - 1^2 = 4$. However, the graph has no low point, as it just keeps on going down forever. Again, the problem is that the graph is not pinned down at the right-hand endpoint, because there is no right-hand endpoint—the interval extends to the right forever.

Suppose you have a continuous function f on a closed and bounded interval $[a, b]$, so that f does have maximum and minimum values on $[a, b]$. How can you find these values? You can't evaluate f at every point x in $[a, b]$, because there are infinitely many such x. The idea is that you can narrow down the possible points x on the x-axis where f *might* have an extreme value to (usually) just a few possibilities. You can then evaluate f at these few possibilities, and pick out the smallest and largest values.

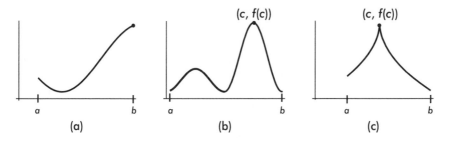

Figure 6.2

Figure 6.2 illustrates the possibilities for those points x at which the function might have a maximum value. Similar remarks apply for a minimum value. In (a), a maximum value occurs at an endpoint of the interval. In (b) and (c), the maximum value occurs at a point $x = c$ inside the interval, that is, at a point c with $a < c < b$. The difference between the two pictures is that in (b), the graph is smooth and rounded at the high point, and has a horizontal tangent line there. This means that f is differentiable at $x = c$, and furthermore that $f'(c) = 0$. In (c), the graph comes to a sharp point at $x = c$. This means, as discussed in Section 3.5, that f *is not differentiable at* $x = c$. These are the only possibilities.

The previous remarks are summarized in the following theorem.

THEOREM 6.1. *Let $f(x)$ be continuous on the closed, bounded interval $[a, b]$. If f has an extreme value at $x = c$ in the interval, then either*

1. *The point c is an endpoint of the interval, so that $c = a$ or b.*
2. *The point c is an inside point and $f'(c) = 0$.*
3. *The point c is an inside point, and f is not differentiable at $x = c$, so that f' is undefined at $x = c$.*

How can you find those points c strictly between a and b where the derivative f' equals zero or is undefined? To find the points where the derivative equals zero, factor the formula for $f'(x)$ as much as you can, and remember that a product equals zero precisely when one of the factors equals zero. Often, the points where the derivative is undefined will show up as follows: the formula for the derivative $f'(x)$ will be a fractional expression, which will be undefined when the denominator equals zero. To spot all of these points, be aware that negative exponents really are the same as fractional expressions, and convert anything with negative exponents to fractional form. In this case, remember also that when the numerator of the fractional expression equals zero, the derivative will be zero.

A Theoretical Aside: Before turning to examples, let's revisit the second case in Theorem 6.1. It was based on the observation that if f had an extreme value at a point c strictly between a and b, and if the graph of f were smooth and rounded at that point, then the tangent line would have to be horizontal. More formally, this means that if f has an extreme value at the inside point $x = c$, and if f is differentiable at $x = c$, then $f'(c) = 0$. It is instructive, if you are so inclined, to see how this can be proven without recourse to any geometric intuition. If you are not so inclined, just skip the next proof.

THEOREM 6.2. *Let $f(x)$ be continuous on the closed, bounded interval $[a, b]$. If f has an extreme value at a point c strictly between a and b, and if f is differentiable at $x = c$, then $f'(c) = 0$.*

Proof: The proof below works when f has a minimum value at c. A similar proof works for the maximum value case.

$$f'(c) = \lim_{h \to 0} \frac{f(c + h) - f(c)}{h}$$

You'll recall from your work with limits that for the limit to exist, the quantity

$$\frac{f(c + h) - f(c)}{h}$$

must approach the *same* limiting value L whether h gets close to 0 from the right (that is, through positive values) or from the left (that is, through negative values). Analyze the sign of $\dfrac{f(c + h) - f(c)}{h}$, keeping in mind that f has a minimum value at c, so that $f(c) \leq f(c + h)$. It follows that the numerator is always positive, but the denominator h is positive only to the right of 0, and is negative to the left of 0. Thus $\dfrac{f(c+h) - f(c)}{h}$ is positive for $h > 0$, but negative for $h < 0$. As $h \to 0$ from the right, the positive numbers $\dfrac{f(c+h) - f(c)}{h}$ get closer and closer to L. As $h \to 0$ from the left, the negative numbers $\dfrac{f(c+h) - f(c)}{h}$ also get closer and closer to the same L. There is only one number L that both positive and negative numbers can get arbitrarily close to at the same time, and that is the number $L = 0$. Hence $f'(c) = 0$.

To review, the points in $[a, b]$ where a continuous f might have an extreme value are the endpoints a and b, or points strictly between a and b where the derivative equals zero or is undefined. Once you have found these points, just evaluate f there and pick out the largest (maximum) value and the smallest (minimum) value.

Example 6.1 Find the maximum and minimum values of $f(x) = 2x^3 - 9x^2$ on the interval $[-1, 5]$.

Solution:

1. First evalute f at the endpoints, to get

$$f(-1) = -2 - 9 = -11 \text{ and } f(5) = 250 - 225 = 25$$

2. Next, compute the derivative $f'(x) = 6x^2 - 18x = 6x(x - 3)$. Now $f'(x)$ equals 0 when either $6x = 0$, that is, when $x = 0$, or when $x - 3 = 0$, that is, when $x = 3$. Evaluating f at these points, you get

$$f(0) = 0 \text{ and } f(3) = 54 - 81 = -27$$

3. The derivative f' is always defined, so it is never undefined.

Of the four values -11, 25, 0, and -27, clearly 25 is the largest and -27 is the smallest. Thus f has a maximum value of 25 at $x = 5$, an endpoint, and a minumum value of -27 at $x = 3$, an inside point.

Example 6.2 For the same function $f(x) = 2x^3 - 9x^2$, find the maximum and minimum values on the interval $[-1, 2]$.

Solution: The important thing for you to notice in this example is that since the interval is $[-1, 2]$, you should only consider points in this interval.

1. First evaluate f at the endpoints, to get

$$f(-1) = -11 \text{ as above, and } f(2) = 16 - 36 = -20$$

2. The only point in the interval where $f' = 0$ is the point $x = 0$, and $f(0) = 0$.
3. The derivative f' is never undefined.

Thus you have the three values -11, -20 and 0 to compare. Clearly the largest is 0, so f has a maximum value on the interval $[-1, 2]$ of 0 at $x = 0$, an inside point, and a minimum value on the interval of -20 at $x = 2$, an endpoint. Although the value of f at the other point, $x = 3$, where $f' = 0$ is -27, a value smaller than -20, the point $x = 3$ is outside the interval you are considering, and so the value of f there doesn't matter.

Example 6.3 Find the maximum and minimum values of $f(x) = x^{\frac{2}{3}}$ on the interval $[-1, 1]$.

Solution:

1. First evaluate f at the endpoints, to get

$$f(-1) = ((-1)^{\frac{1}{3}})^2 = (-1)^2 = 1, \text{ and } f(1) = 1$$

2. $f'(x) = \frac{2}{3}x^{-\frac{1}{3}} = \frac{2}{3x^{\frac{1}{3}}}$. This fractional quantity is *never* 0 because the numerator is never 0.
3. However, $f'(x)$ is undefined at the point $x = 0$, where the denominator equals 0. The value of f there is $f(0) = 0$.

It is clear that the minimum value of 0 occurs at the inside point $x = 0$ and the maximum value of 1 occurs at the endpoints $x = -1$ and $x = 1$. The graph of f on the interval $[-1, 1]$ is pictured in Figure 6.3.

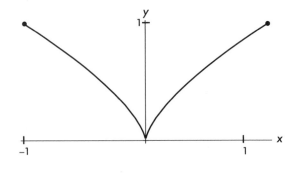

Figure 6.3. $y = x^{\frac{2}{3}}$ on $[-1, 1]$

Local or Relative Extreme Values

In addition to the points where a function might have a maximum or minimum value, there are other points that are important for the behavior of the function and the shape of its graph. Look at the curve in Figure 6.4.

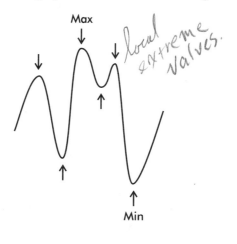

Figure 6.4

Its high point, or maximum, and its low point, or minimum, are indicated in the picture. If you thought of the graph as the profile of a landscape, the maximum could represent the highest hill in the landscape, while the minimum could represent the deepest valley. The other points indicated, which look like tops of hills (although not the biggest hills), and bottoms of valleys (although not the deepest valleys), are also important for understanding a curve, and are called **relative** or **local extreme values**, to distinguish them from the extreme values on the whole interval, which are called **absolute** or **global extreme values**. You can see that what distinguishes the top of a hill is that it is higher than *nearby* points, and likewise what distinguishes the bottom of a valley is that it is lower than *nearby* points. A formal definition is given by the following:

DEFINITION 6.2. *A function $f(x)$ has a **local maximum** at the point $(c, f(c))$, if there is some interval about c such that $f(c) \geq f(x)$ for all x in that interval. A function $f(x)$ has a **local minimum** at a point c, if there is some interval about c such that $f(c) \leq f(x)$ for all x in that interval.*

It is important to realize that the phrase "some interval about c" means some interval that extends both to the left of $x = c$ and to the right of $x = c$.

Just as with global extreme values, derivatives can help you identify local extreme values.

THEOREM 6.3. *If f has a local extreme value at $(c, f(c))$ and is differentiable at that point c, then $f'(c) = 0$.*

Proof: The proof is *exactly* the same as that of Theorem 6.2. If you look at that proof, you will see that it only involved values of f at points $c + h$ for h close to 0, that is, it only involved points $c + h$ close to c.

It follows that if f has a local extreme at $(c, f(c))$, then either f is not differentiable at $x = c$, or if it is differentiable, then $f'(c) = 0$. Thus, to find local extreme values, just as you did for global extreme values, you look for points where the derivative is either undefined or equal to zero. Such points have a special name.

DEFINITION 6.3. *Let f be a function. If f is defined at the point $x = c$ and either*

$$f'(c) = 0 \ or$$

$$f'(c) \ is \ undefined$$

then the point c is called a ***critical point*** *of f.*

6.2 THE MEAN VALUE THEOREM

You learned in Theorem 5.1 that if $f(x)$ is a constant function, then its derivative equals zero. The converse *seems* obvious—namely, that if the derivative $f'(x) = 0$ *on an interval*, then $f(x)$ must be constant on that interval. After all, the derivative means the rate of change, and if the rate of change of f equals zero, then f doesn't change, so it *must* be a constant. Even though this *seems* obvious, the *proof* of this fact requires a new theorem. There are several other almost intuitively obvious facts whose proofs also require the same new theorem.

DEFINITION 6.4. *A function $f(x)$ is said to be **strictly increasing on an interval I** if, whenever x_1 and x_2 are any two points in I with $x_1 < x_2$, then $f(x_1) < f(x_2)$.*

The definition means that as the variable gets larger, so do the values of the function, so that as you go from left to right on the x-axis, the graph goes up. It is not too hard to see that if f is increasing on an interval I, and f is differentiable on I, then $f'(x) \geq 0$ for all x in I. One way to see this is graphically: if the curve is increasing, then the tangent line segments also rise as you go from left to right along the x-axis, and thus have positive slope. But the slope of the tangent line to the graph of f at $(x, f(x))$ is just $f'(x)$ (see Figure 6.5).

A Theoretical Aside: You can *prove* the above fact using the definition of the derivative. Recall that:

$$f'(x) = \lim_{h \to 0} \frac{f(x + h) - f(x)}{h}$$

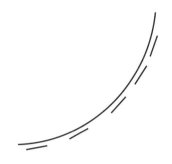

Figure 6.5. *An increasing function*

Now for h close to 0 but positive, $x + h > h$, and as f is increasing, it follows that $f(x+h) > f(x)$, so that both the numerator and denominator are positive. Thus $\dfrac{f(x+h) - f(x)}{h}$ is also positive, and getting closer and closer to $f'(x)$ as h gets close to 0. Positive numbers can get arbitrarily close to only another positive number or 0, which gives $f'(x) \geq 0$.

Now the converse fact, that if $f'(x) > 0$ for all x in some interval I, then f is strictly increasing on I, also seems obvious. All the line segments tangent to the curve have positive slope, so they go up as you go horizontally from left to right. This seems to force the curve to do the same. The idea is basically correct, but the proof requires a *new* theorem.

DEFINITION 6.5. *A function $f(x)$ is said to be **strictly decreasing on an interval I** if, whenever x_1 and x_2 are any two points in I with $x_1 < x_2$, then $f(x_1) > f(x_2)$.*

All of the above remarks about increasing functions and positive derivatives hold true for decreasing functions and negative derivatives. Thus you can show from the definition of the derivative that if a function is decreasing on an interval then its derivative must be less than or equal to 0 on the interval. Conversely, it seems obvious that if $f'(x) < 0$ on an interval, then f must be decreasing. As with functions with 0 derivative or positive derivative, this requires a *new* theorem, which follows.

THEOREM 6.4. *(The Mean Value Theorem—MVT) Let the function $f(x)$ be continuous on the closed, bounded interval $[a, b]$, and differentiable at every point strictly between a and b. Then there exists some point $x = c$ (and maybe more than one) strictly between a and b, such that*

$$\frac{f(b) - f(a)}{b - a} = f'(c)$$

What the theorem states is that at some point between a and b, the tangent line to the graph has the same slope as the line segment connecting the endpoints of the graph. In Figure 6.6 two such points are indicated.

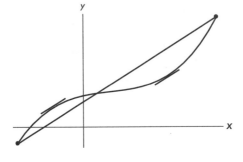

Figure 6.6

Proof: The Mean Value Theorem is proven by first verifying a special case, called **Rolle's Theorem**. Rolle's Theorem covers the special case when $f(a) = f(b) = 0$. Of course then $\dfrac{f(b) - f(a)}{b - a} = 0$, so Rolle's Theorem states that in this case, there is some point c strictly between a and b with $f'(c) = 0$. To see why this is so, recall from Section 6.1 that the continuous function f has a maximum and minimum value on $[a, b]$. If both these extreme values occur at the endpoints where f has the value 0, then f is constant and equal to 0 on $[a, b]$, and $f' = 0$ *everywhere* in $[a, b]$. If one of the extreme values occurs at some point c strictly between a and b, then by Theorem 6.2, $f'(c) = 0$.

The general case follows from the special case by applying Rolle's Theorem to the function $f(x) - h(x)$, where $h(x)$ is the formula for the line connecting the endpoints $(a, f(a))$ and $(b, f(b))$ of the graph. A picture is given in Figure 6.7, and you can find the details of the formula in any standard calculus text. What happens is sort of like this: Try to imagine from the picture that as the line is subtracted, the curve $f(x)$ both rotates and moves down, so that Rolle's theorem applies. Find the point where the new curve has a horizontal tangent line. When you raise and rotate the curve to its original position, the formerly horizontal tangent line will have slope equal to that of the line connecting the endpoints of the graph.

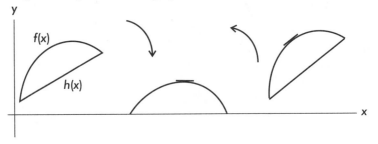

Figure 6.7

Example 6.4 For the function $f(x) = x^3 - 3x$ on the interval $[1, 4]$, find the point or points c predicted by the Mean Value Theorem.

Solution:

$$\frac{f(b) - f(a)}{b - a} = \frac{f(4) - f(1)}{4 - 1}$$

$$= \frac{(64 - 12) - (1 - 3)}{4 - 1}$$

$$= \frac{52 - (-2)}{3} = \frac{54}{3} = 18$$

As $f'(x) = 3x^2 - 3$, you are looking for a point c in $[1, 4]$ satisfying the equation $3x^2 - 3 = 18$, or equivalently, $x^2 = \frac{21}{3} = 7$. The points satisfying this last equation are $\pm\sqrt{7} \approx \pm 2.65$. However, only the point $+\sqrt{7} \approx 2.65$ is the correct answer to the problem, as the Mean Value Theorem talks about points c *inside* the interval $[a, b]$, and the point $-\sqrt{7} \approx -2.65$ is outside the interval $[1, 4]$ of this example.

Using the Mean Value Theorem, you can prove the intuitively obvious facts mentioned in the first part of this section.

THEOREM 6.5. *If f is differentiable on an interval and the derivative $f'(x) = 0$ for all x in the interval, then f is constant on the interval.*

Proof: It is enough to take any two points q and r on the interval, with $q < r$, and show that $f(q) = f(r)$. Let q and r play the role of a and b in the statement of the Mean Value Theorem. There is a point c strictly between q and r with

$$\frac{f(r) - f(q)}{r - q} = f'(c)$$

As $f'(c) = 0$ by hypothesis (since c is a point in the interval), it follows that the fractional expression $\dfrac{f(r) - f(q)}{r - q} = 0$. The only way a fraction can equal 0 is when the numerator equals 0, so $f(r) = f(q)$.

THEOREM 6.6. *If f is differentiable on an interval and $f'(x) > 0$ for all points x in the interval, then f is strictly increasing on the interval.*

Proof: Again, let q and r be any two points in the interval, with $q < r$. To show that f is strictly increasing, you must show that $f(q) < f(r)$. Apply the Mean Value Theorem to the smaller interval $[q, r]$, by letting q and r play the role of a and b in the statement of the Mean Value Theorem. There is at least one point c strictly between q and r such that

$$\frac{f(r) - f(q)}{r - q} = f'(c)$$

Now $f'(c) > 0$, so $\dfrac{f(r) - f(q)}{r - q} > 0$ also. As $r - q > 0$ (since $q < r$), it follows that $f(r) - f(q) > 0$, that is, that $f(r) > f(q)$ or $f(q) < f(r)$.

THEOREM 6.7. *If f is differentiable on an interval and $f'(x) < 0$ for all points x in the interval, then f is strictly decreasing on the interval.*

Proof: The proof is similar, and we will skip it.

6.3 CURVE SKETCHING

The topic of using calculus to sketch the graphs of curves may be a little outmoded in the age of graphing calculators, but it is a classical and famous type of application. It will give you good practice in seeing how much extra information you can get about a function from its derivative, information that is not so directly accessible from the formula for the function itself.

In sketching a graph, you would want to know for what intervals the curve is increasing or rising, and for what intervals it is decreasing or falling. From Theorem 6.6 and Theorem 6.7, this means you have to find the intervals on which the function has positive derivative, and the intervals on which the function has negative derivative. How do you find these intervals? As so often happens in mathematics, you find the answer to a question by first answering a different, or even opposite, question. If the derivative is not positive or negative, what could it be? The only other possibilities are *zero* or *undefined*.

The technique is to look for the points where the derivative is zero or undefined, and think of those points as dividing the x-axis into intervals. On each interval, the derivative will be either always positive, or always negative. How do you know this? Well, the derivative won't be zero or undefined in an interval because each interval is *between* the points where the derivative is zero or undefined. Thus, at each point of an interval the derivative will be positive or negative, but never zero or undefined. It seems unlikely that the derivative could jump from being positive to being negative, inside an interval, without also being zero or undefined somewhere in the interval, which it can't be. It is so unlikely that in fact it does not happen.

If you know that in a given interval, the derivative is *always* positive or *always* negative, the easiest way to figure out which is to pick a *sample point* in the interval, and evaluate the derivative there. The sign of the derivative at the sample point tells you the sign of the derivative on the whole interval, because on each interval the derivative is always either positive or negative, one or the other, but not both.

When you find a point $x = c$ where the derivative $f'(c)$ either equals zero or is undefined, remember from Definition 6.3 that if the function f itself is defined at $x = c$, then $x = c$ is a *critical point* for f. A look at Figure 6.4 should convince you that it is quite likely (although not definite) that the point $(c, f(c))$ on the graph might be the "top of a hill" or the "bottom of a valley," that is, a local extreme, and thus important in determining the shape of the graph. Thus you should always plot these points on the graph of f. Remember also in graphing such points that when $f'(c) = 0$ the top of the hill or bottom of the valley is smooth and rounded, while when $f'(c)$ is undefined, the top of the hill or bottom of the valley comes to a sharp point (see Figures 6.2(b) and 6.2(c)).

It is important for you to develop some system for organizing your work, and for correctly carrying the results of your work over to the graph. You'll see the system I like best below, but if you want to use a different system, that's fine. Just find some system to organize your work. I will first work through one example, explaining the steps as I go along, and then just follow these steps for another example.

Example 6.5 Use derivatives to sketch the graph of $f(x) = x^3 - 3x^2$.

Solution:

1. Compute the derivative, and find the points where $f'(x)$ equals 0 or is undefined.

   (a) Remember, the standard general way of finding where $f'(x) = 0$ is to factor the formula for f' as much as you can, and then set each factor equal to 0.

   (b) If f' is a fractional expression, it will be zero when the numerator equals 0, and undefined when the denominator equals 0.

   (c) Remember to convert negative exponents in the formula for f' into fractional expressions.

   In this example, $f'(x) = 3x^2 - 6x = 3x(x - 2)$, which equals 0 when either $3x = 0$ or when $x - 2 = 0$. Thus $f'(x) = 0$ for $x = 0$ or 2. The derivative is never undefined in this example.

2. In your sketch, mark off with some symbol (I use an X) the points 0 and 2 on the x-axis (see Figure 6.8). Remember, these are the points on the x-axis where f' behaves a certain way. They are not necessarily going to be points that are also on the graph of f. Such a point will be on the graph of f only when the value of f at the point equals 0.

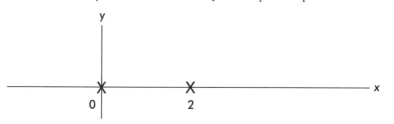

Figure 6.8

3. The two points $x = 0$ and $x = 2$ divide the line into *three* intervals (look at the sketch) on each of which the derivative is either positive or negative. Pick a sample point in each interval and evaluate f' there to figure out which. I organize my work in a table.

Interval	Sample point	f'(sample point)	Sign of f'	Graph of f is
$(-\infty, 0)$	-1	$3(-1)^2 - 6(-1) = 3 + 6 = 9$	$+$	increasing
$(0, 2)$	1	$3(1)^2 - 6 \cdot 1 = 3 - 6 = -3$	$-$	decreasing
$(2, \infty)$	3	$3(3^2) - 6 \cdot 3 = 27 - 18 = 9$	$+$	increasing

4. Transfer the information about whether the curve is increasing or decreasing on the intervals over to the sketch. I do this with arrows in Figure 6.9.

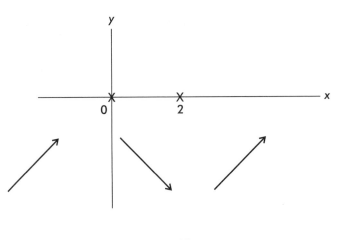

Figure 6.9

5. Evaluate f at the points you found in Step 1, provided the function f is defined at these points. That is, if these points are critical points, evaluate f and get points to plot on the graph of f. They are likely to be important for the sketch.

$$f(0) = 0^3 - 3 \cdot 0^2 = 0, \ f(2) = 2^3 - 3 \cdot 2^2 = 8 - 12 = -4$$

6. Plot these values on the sketch of the graph. Check your work for consistency. Remember, since you already found that f was decreasing on the interval $(0, 2)$, for consistency you must have $f(2) < f(0)$.

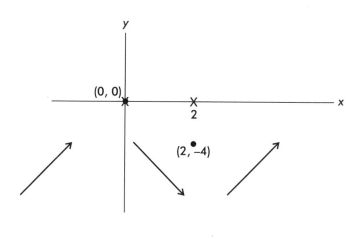

Figure 6.10

7. Do your best to fill in the rest of the graph, in a fashion consistent with what you have already determined.

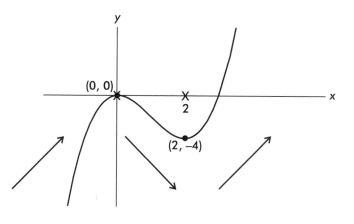

Figure 6.11

Notice that the two *critical* points you found in Step 1 turn out to be local extreme values. The function has no global extreme values.

Example 6.6 Use derivatives to sketch the graph of $f(x) = x^{\frac{2}{3}}$.

Solution:

1. $f'(x) = \frac{2}{3}x^{-\frac{1}{3}} = \frac{2}{3x^{\frac{1}{3}}}$. As the numerator is never 0, the derivative is never 0, but it is undefined when the denominator equals 0, at $x = 0$.

2. Mark off this point $x = 0$ on the x-axis.

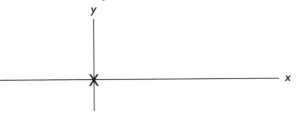

Figure 6.12

3. The one point $x = 0$ divides the line into *two* intervals, on each of which f' is positive or is negative. Figure out which by the sample point method.

Interval	Sample point	f′(sample point)	Sign of f′	Graph of f is
$(-\infty, 0)$	-1	$\frac{2}{3 \cdot -1^{\frac{1}{3}}} = \frac{2}{-3}$	$-$	decreasing
$(0, \infty)$	1	$\frac{2}{3 \cdot 1^{\frac{1}{3}}} = \frac{2}{3}$	$+$	increasing

4. Transfer this information to the graph.

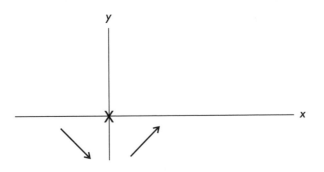

Figure 6.13

5. $f(0) = 0^{\frac{2}{3}} = 0$
6. Transfer this information to the graph.

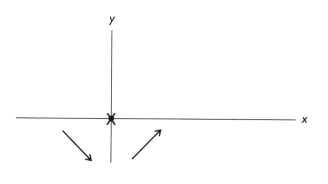

Figure 6.14

7. In sketching, remember that while f is defined at 0, f' is *undefined* at $x = 0$. This generally has the effect of giving the graph of f a sharp point at $x = 0$.

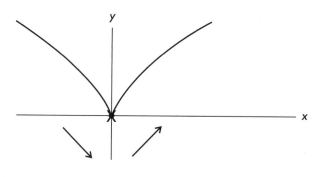

Figure 6.15

In this final example, I will skip all the intermediate graphs, and just go through the steps.

Example 6.7 Use derivatives to sketch the graph of $f(x) = \dfrac{1}{x - 2}$.

1. Use the quotient rule to compute the derivative of f:

$$f'(x) = \frac{(x - 2) \cdot 1' - 1 \cdot (x - 2)'}{(x - 2)^2}$$

$$= \frac{0 - 1}{(x - 2)^2}$$

$$= \frac{-1}{(x-2)^2}$$

As the numerator is never 0, f' never equals 0. However, f' is undefined at the point $x = 2$, where the denominator equals 0.

2. Mark off the point $x = 2$ with an X on the x-axis.

3. The point $x = 2$ divides the line into two intervals, on each of which f' is either always positive or always negative. Figure out which by the sample point method.

Interval	Sample point	f'(sample point)	Sign of f'	Graph of f is
$(-\infty, 2)$	0	$\dfrac{-1}{(-2)^2}$	$-$	decreasing
$(2, \infty)$	3	$\dfrac{-1}{1^2}$	$-$	decreasing

4. Transfer this information to the graph.

5. You cannot compute $f(2)$ because the function f itself is undefined at 2. Thus, at $x = 2$, there is no point on the graph. However, the fact that substituting $x = 2$ into the formula for f would give you $\dfrac{1}{0}$ means, as in Section 3.3, that as x gets close to 2, from the left or the right, the values of $f(x)$ either get arbitrarily large, or arbitrarily large in the negative direction. The information in the table on whether f is increasing or decreasing tells you which.

6. There is no value to plot at $x = 2$.

7. Sketch the graph (see Figure 6.16). Remember, how the graph of f behaves near $x = 2$ should be consistent with the arrows showing whether f is increasing or decreasing.

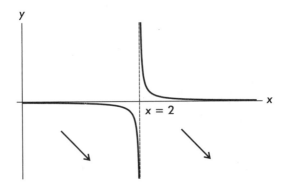

Figure 6.16

In a standard calculus course, you would additionally learn about $\lim_{x \to \pm\infty}$ $f(x)$, horizontal and vertical asymptotes, and the use of the *second derivative*, that is, the derivative of the derivative, to give you finer information about the shape of the curve. The information the second derivative gives you should be organized in much the same way that you organized the first derivative information.

6.4 SUMMARY OF MAIN POINTS

- For a continuous function f on a closed and bounded interval $[a, b]$, the possible points x at which f might have an extreme value are the endpoints a and b, and the critical points where either the derivative $f'(x)$ equals zero or is undefined.

- Evaluate f at the endpoints and at the critical points to determine the maximum and minimum values.

- In finding the critical points, remember that if the formula for $f'(x)$ has terms with negative exponents, convert them to fractional form. A fractional expression generally equals zero when its numerator equals zero, and is undefined when its denominator equals zero.

- The general method for determining when an expression equals zero is to try to factor it as much as possible, and then set each factor equal to zero.

- (The Mean Value Theorem) If $f(x)$ is continuous on the closed and bounded interval $[a, b]$ and differentiable on the open interval (a, b), then there is *at least* one point c strictly between a and b such that

$$f'(c) = \frac{f(b) - f(a)}{b - a}$$

- The Mean Value Theorem can be interpreted as stating that there is at least one point strictly between a and b where the tangent line has the same slope as the line connecting the endpoints $(a, f(a))$ and $(b, f(b))$ of the graph.

- It is because of the Mean Value Theorem that you can prove the following:

  — If $f'(x) = 0$ on an interval, then f is constant on that interval.

  — If $f'(x) > 0$ on an interval, then f is strictly increasing on that interval.

  — If $f'(x) < 0$ on an interval, then f is strictly decreasing on that interval.

- To make use of the first derivative in curve sketching:
    — Find the critical points where $f'(x)$ equals 0 or is undefined.
    — Use these points to mark off intervals on the line.
    — On each of these intervals, $f'(x)$ is either strictly positive or strictly negative. Figure out which by evaluating f' at a sample point in each interval.
    — Remember that if $f' > 0$ then f is increasing, and if $f' < 0$ then f is decreasing. Mark your graph accordingly.
    — Evaluate f at the critical points, if possible, and plot these on the graph.
    — Sketch the curve in a manner consistent with the information you have obtained so far.

6.5 EXERCISES

For the Exercises 1–7, find the maximum and minimum values of the given functions on the indicated interval.

1. $f(x) = x^3 - 3x^2 - 9x + 5$ on the interval $[0, 4]$
2. $f(x) = x^3 - 3x^2 - 9x + 5$ on the interval $[-2, 2]$
3. $f(x) = 6x^2 - 4x$ on the interval $[0, 1]$
4. $f(x) = 6x^2 - 4x$ on the interval $[2, 4]$
5. $f(x) = \dfrac{1}{4}x - \sqrt{x}$ on the interval $[1, 25]$
6. $f(x) = \dfrac{1}{4}x - \sqrt{x}$ on the interval $[0, 9]$
7. $f(x) = \dfrac{1}{4}x - \sqrt{x}$ on the interval $[0, 1]$

For the Exercises 8–12, and for the given function on the indicated interval, find the point or points c predicted by the Mean Value Theorem.

8. $f(x) = x^2 - 4x + 3$ on the interval $[0, 4]$
9. $f(x) = x^2 - 4x + 3$ on the interval $[-2, 4]$
10. $f(x) = x^3 - 2x + 1$ on the interval $[-2, 3]$
11. $f(x) = x^3 - 2x + 1$ on the interval $[-1, 3]$
12. $f(x) = x + \dfrac{1}{x}$ on the interval $[1, 2]$

For the functions in Exercises 13–16, use the first derivative to sketch the graph. Be sure to indicate clearly the intervals on which the function is increasing, the intervals on which the function is decreasing, and also all local maximum values and local minimum values.

13. $f(x) = x^2 - 4x + 8$

14. $f(x) = x^3 - 6x^2 + 9x$

15. $f(x) = x^4 - 4x^3 + 1$

16. $f(x) = x + 3x^{\frac{2}{3}}$

7
WORD PROBLEMS

Word problems are the shoals upon which many troubled calculus students founder. The hardest part of a word problem is the reading of the English and the setting up of the formulas. Applying calculus to the formulas to solve the problem is not as difficult. In this section you will find step-by-step guidance on how to approach the two main types of word problems, maximum-minimum problems (max-min problems, for short), and related rate problems. The guidance will include a brief review of basic geometry, and some hints in reading comprehension. But be warned—no one can give you a mechanical list of steps to follow to ensure that you can solve every word problem, only a list of steps to serve as a rough guide. Each problem is different, and furthermore the same problem can be phrased in a variety of ways. English is a very flexible language. The most important skill in solving word problems is *reading comprehension*. The most important attitude to have in attacking word problems is to be willing to think about what you are reading—give up on hoping to *mechanically* apply a set of steps.

7.1 A REVIEW OF GEOMETRY

One thing you learn in geometry is a collection of basic formulas for length, area, and volume. In calculus, students sometimes get length, area, and volume (or the formulas for these quantities) mixed up. Here is an easy way to tell your length formulas from your area formulas from your volume formulas: Most basic shapes are determined by, or described in terms of, some fundamental measures of length. For example, a square is determined by the length of an edge; a rectangle by its length and width; a circle or a sphere by its radius; a box by its length, width, and height; and both a right circular cylinder and right circular cone (the only kinds of cylinders and cones you will see in this book) by the base radius and the height. Now length is one-dimensional, area is two-dimensional, and volume is three-dimensional. Because of this, all length formulas should in essence involve quantities of length taken to the first power only, while formulas for area should involve either quantities of length taken to the second power, or two different quantities of length multiplied together. In other words, for area formulas, if you add up the powers of all length terms that are multiplied together, you should get 2. As volume is three-dimensional, if you add up all the powers of length terms that are multiplied together, you should get 3.

Thus, the formula for the volume of a box is $V = lwh$. Each of l, w, and h represents a different length associated with a box, and each occurs to the first power. Adding up one power for l, one for w, and one for h gives 3—just right for the three-dimensional quantity of volume. A term like l^2h or x^2y or lwh should *not* appear in a formula for the surface area of a box. As area is two-dimensional, an area formula should have only the square of certain length quantities, or two different length quantities multiplied together. If you do this quick dimensional analysis for every general formula you write down, you will never make the mistake of writing r^2h for the curved surface area of a cylinder, or $2\pi rh$ for the volume. A term such as r^2h has to be part of a volume formula, and a term such as $2\pi rh$ has to be part of an area formula.

In connecting the English words to the formulas, remember that areas are given in terms of units of length squared, such as 3 square feet, written 3 ft^2, or 9.4 square meters, written 9.4 m^2, while volumes are given in terms of units of length cubed, such as 3 cubic feet (3 ft^3).

Now a square is determined by its edge length x, and most formulas for squares are in terms of edge length. Thus the area $A = x^2$, and the perimeter $P = 4x$. Most students are used to seeing a formula like $A = x^2$ with problems such as: "If a square has edge length 5 inches, what is its area?" You substitute $x = 5$ into the right-hand side, and compute $A = 5^2 = 25$ square inches. You should view formulas more flexibly. The formula $A = x^2$ expresses a relationship between x and A, and if you know one, you can get the other. In other words, you should be able to use a formula from left to right, as well as from right to left. Thus, if you know that the area of a square is 25 square inches, then use $A = x^2$ by substituting 25 for A, so that $25 = x^2$, to get $x = 5$. Furthermore, as area and edge length are linked, and perimeter and edge length are linked, then area and perimeter are linked (with edge length being the link).

Example 7.1 Find a formula for the area of a square as a function of its perimeter P.

> *Solution:* You must use the link x of edge length. If you know the perimeter P, what can you do to get the area A? First of all, if you know the perimeter P, you can get the edge length x by solving $P = 4x$ for x, so that $x = \dfrac{P}{4}$. Now that you know x, you can get A using $A = x^2 = \left(\dfrac{P}{4}\right)^2 = \dfrac{P^2}{16}$.

Rectangles are determined by their lengths and widths, with area being given by $A = lw$ and perimeter by $P = 2l + 2w$. For a rectangle, if you know only the area you cannot determine anything else. For example, a 20 ft^2 rectangle could be 5 by 4, or 10 by 2. The 5-by-4 rectangle has perimeter 18, while the 10-by-2 has perimeter 24.

The Pythagorean theorem gives the hypotenuse c of a right triangle in terms of the short sides a and b by the formula $c = \sqrt{a^2 + b^2}$. Now c is a length, and you might think that this formula violates what you read above—namely, that length is one-dimensional, so length formulas should involve only length quantities to the first power. The Pythagorean theorem has length quantities to the second power. However, the $\sqrt{}$ serves to convert the second power back to the first power, so that the dimensional analysis works here, too.

The area of a circle of radius r is πr^2, and its circumference is $2\pi r$. The volume of a sphere of radius r is $\frac{4}{3}\pi r^3$, and its surface area is $4\pi r^2$.

Right circular cylinders are determined by their base radius r and their height h. All the cross sections perpendicular to the height are circles of the same area, πr^2 (see Figure 7.1).

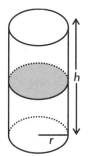

Figure 7.1. *A right circular cylinder*

It is a general geometric principle that if you have a volume, and all cross sections perpendicular to some direction have the same area A, then the volume $V = A \cdot h$, where h is the extent of the figure in the given direction. Thus the volume V of a cylinder equals $\pi r^2 h$. The curved surface area of a cylinder is $2\pi r h$. To see why, imagine taking a cylinder without a top or bottom, cutting it as shown in Figure 7.2, and spreading it out to get a rectangle. If you have a right circular cylinder with both a bottom and a top, it is easy to get the total surface area. You know now the formula for the curved surface area, and the bottom and top are each circles of radius r, so the total surface area is given by $S = 2\pi r h + 2\pi r^2$.

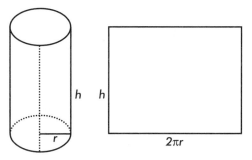

Figure 7.2

A right circular cone is also determined by its base radius r and its height h (see Figure 7.3). The formula for the volume of the cone happens to be $V = \frac{1}{3}\pi r^2 h$. It is hard to justify this formula by elementary mathematics.

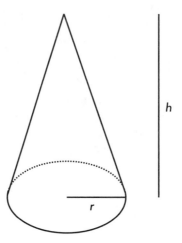

Figure 7.3

Finally, a box is determined by its length l, width w, and height h (see Figure 7.4).

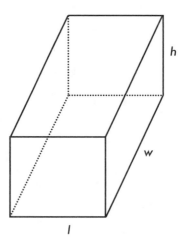

Figure 7.4

The volume of a box is $V = lwh$. You should know how to find the total surface area of a box. The surfaces are all rectangles, and you know how to find the area of a rectangle. Just observe that the surfaces come in pairs: front-back, right-left, and bottom-top. The front in Figure 7.4 is an $l \times h$ rectangle, and so is the back. The right is a $w \times h$ rectangle, and so is the left. Finally, the bottom is an $l \times w$ rectangle, and so is the top. Thus the total surface area of a box is $S = 2lh + 2wh + 2lw$.

Be sure to read all problems carefully. For example, if a problem talks about a box with an open top, then there is no top, and there are only five surfaces to consider, giving a surface area formula of $S = 2lh + 2wh + lw$. If a problem talks about a box with a square base, then l and w are equal, and the formulas for volume and surface area involve only two distinct variables, not three.

In many geometry problems, even after you draw the picture, the relationship between quantities is not clear until you cleverly draw in just the right extra line. Once you see this done, you should remember it. Below are some examples of the extra-line trick.

Example 7.2 Car A is 20 feet east of an intersection, and car B is 30 feet north of the intersection. How far apart are they?

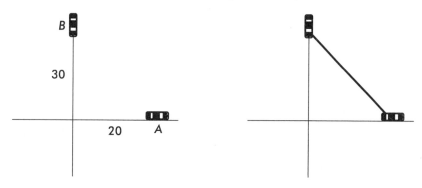

Figure 7.5

Solution: On the left of Figure 7.5 is the picture of the situation, and on the right is the picture with the line representing the distance between the cars explicitly drawn in. Now it is clear from the diagram that you should use the Pythagorean theorem to get

$$\text{distance} = \sqrt{20^2 + 30^2} = \sqrt{1300} \approx 36.06 \text{ feet}$$

Example 7.3 An inverted conical paper cup has a circular top of radius 2 inches, and a depth of 6 inches. It is filled to a depth of 3 inches with water. What is the volume of the water in the cup?

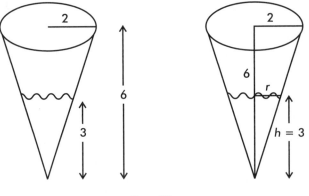

Figure 7.6

Solution: On the left of Figure 7.6 is the picture of the situation, and on the right a somewhat different picture, with an extra line drawn in, that should make you see similar triangles. Thus,

$$\frac{6}{2} = \frac{h}{r} = \frac{3}{r}$$

so that $r = 1$. Thus the water itself forms a cone of radius 1 and height 3, so the volume of the water equals $V = \frac{1}{3}\pi 1^2 \cdot 3 = \pi$ cubic inches.

Example 7.4 A rectangle is inscribed in a circle of radius 4. If the length of the rectangle is 5, what is its width?

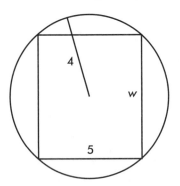 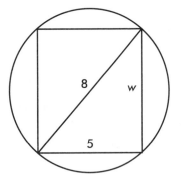

Figure 7.7

Solution: On the left of Figure 7.7 is the picture of the described situation, and on the right the same picture with the confusing radius line omitted, and the diagonal of the rectangle drawn in instead. Once you notice that the diagonal is the diameter $2r = 8$, you can apply the Pythagorean theorem to get

$$8^2 = 5^2 + w^2$$

Thus, $w = \sqrt{64 - 25} = \sqrt{39} \approx 6.24$.

7.2 MAX-MIN WORD PROBLEMS

All max-min word problems ask you to find the largest or the smallest value a function has on an interval. The hard part is usually reading the English and getting the formula for the function. Once you have the function, you learned in Chapter 6 how to find the largest or smallest values. Below is a list of steps to go through to try systematically to solve any such problem. The list should be used to guide you systematically through the process, but it cannot be applied unthinkingly—you really must reason your way through every single word problem. After we discuss the items on the list, you will see some particular examples worked out according to the steps on the list.

1. Read the problem once over quickly.

    If it is a four-sentence problem and you are confused by the second sentence, don't stop there. Just read the whole problem. Maybe the third or fourth sentence will clarify the second sentence.

2. Read the problem again carefully, until you understand what is going on.

    One way to see if you understand the problem is to see if you can rephrase it in your own words.

3. If the problem is a geometry problem, draw a picture and label what there is to label.

    Don't do too much at this stage. If the problem is about a tin can (right circular cylinder), just draw the cylinder and label r and h. Don't worry right now about whatever else they tell you about the can.

4. Identify in words whether you are maximizing or minimizing, and what it is exactly that you are maximizing or minimizing.

    One hint is to look for superlative adjectives—biggest, smallest, largest, cheapest, fastest, most, least—and the nouns they refer to, or superlative adverb phrases—as cheaply as possible, as large as possible—and what these phrases refer to. You might be lucky and actually see the words "maximum" or "minimum" in the problem.

5. Write the *general* formula for what it is you are supposed to maximize or minimize.

Again, don't do too much at this stage. If the problem asks you to find the tin can using the least amount of metal, you want to minimize the surface area of the metal, so just write $S = 2\pi r^2 + 2\pi rh$.

If your formula contains only one variable, skip the next three steps. If your formula contains more than one variable, do them.

6. Look at the problem again carefully to try to get a relationship between the variables. Write this relationship down as a *Given*.

If you want a tin can holding 75 cubic inches, write down: "*Given V =* $75 = \pi r^2 h$."

7. Do the algebra to solve for one variable in the *Given* equation as a function of the other.

Look carefully at the *Given* equation. Although in principle it doesn't matter which of the two variables you solve for, in practice solving for one of the variables instead of the other might be a lot easier. For example, in the given equation above of $75 = \pi r^2 h$, solving for h in terms of r is not too bad. Solving for r in terms of h requires the introduction of square roots.

8. Use your formula for one variable in terms of the other to rewrite the formula for what you want to maximize or minimize as a function of one variable only.

9. Write down the interval over which the above variable can vary, for the particular word problem you are solving.

For example, if x is one side of a rectangle, and you know the perimeter is 400 feet, then $x > 0$ and $x < 200$, so the interval is $(0, 200)$. If you want to, you can often add the endpoints and consider the interval to be $[0, 200]$. This allows you to use the methods of Section 6.1.

10. Take the derivative, and do the algebra to determine where the derivative equals 0 and where it is undefined.

Remember to write negative exponents in fractional form. Remember from Section 6.1 that the extreme values occur at the above critical points (where the derivative equals 0 or is undefined), or possibly at endpoints if your interval has endpoints.

11. If you are working on a closed, bounded interval, use the methods of Section 6.1. If not, you may have to use the methods of Sections 6.2 and 6.3 to identify intervals on which the function is increasing or decreasing, to see which is a maximum value and which is a minimum value.

12. Be sure you answer *exactly* what the problem is asking for.

That was certainly a long list of steps, but after a few problems, it should become automatic as long as you understand what is going on.

Example 7.5 What is the largest possible product you can form from two numbers that are each greater than or equal to 0, and whose sum equals 25?

Solution:

1. Read the *whole* problem a couple of times.
2. You want to pick two nonnegative numbers whose sum is 25, so that the product is as big as possible.
3. This problem is not a geometry problem, so there is no picture.
4. You want to maximize the product. The clue comes from the phrase "largest possible product."
5. This step is sometimes hard for students. How can you write the formula for the product of two numbers? First you have to call the numbers something, such as x and y. Now let P stand for the product, so that

$$P = xy$$

6. The formula contains two variables, so you must find a relationship between them. There has to be such a relationship, or constraint, because otherwise there would be no maximum product. Whatever two numbers x and y you picked, you could get a bigger product by picking bigger numbers x and y. The relationship comes from the words "whose sum equals 25." Thus the *Given* is:

$$x + y = 25$$

7. Now do the algebra. As $x + y = 25$, $y = 25 - x$. In this case, it is equally easy to solve for y in terms of x as the other way around.
8. $P = xy = x(25 - x) = 25x - x^2$
9. As the problem states that x is nonnegative, clearly $x \geq 0$. As $y = 25 - x$ is also ≥ 0, it follows that $25 \geq x$, so that $x \leq 25$. Thus the interval over which x can vary in this problem is $[0, 25]$.

10. $P' = 25 - 2x$. Thus P' is never undefined, but it equals 0 for $x = \dfrac{25}{2} = 12.5$.

11. $P(12.5) = 25 \cdot 12.5 - (12.5)^2 = 156.25$. As you are considering values of the function P for x in the closed and bounded interval $[0, 25]$, the methods of Section 6.1 apply. You must compare the values of P at the endpoints of the interval with the value of P at the critical point. At the endpoints you get $P(0) = P(25) = 0$.

12. The largest possible product you can form is 156.25.

Example 7.6 A farmer has 400 feet of fencing material, and wants to enclose a rectangular pasture. What should the dimensions of the field be in order to enclose the most area?

Solution:

1. Read the *whole* problem a couple of times.
2. Of all the possible rectangles you can make with 400 feet of fencing (see Figure 7.8), which one has the largest area?

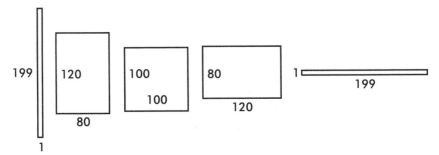

Figure 7.8. *Some of the infinite possibilities*

3. This is clearly a geometry problem. A rectangle is determined most basically by its length and width, so the labeled picture should look like that in Figure 7.9:

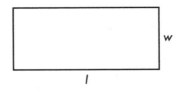

Figure 7.9

4. The problem talks about the "most area," so you want to maximize area.
5. $A = lw$
6. Again, you could enclose more area by enlarging the length and width, if it were not for the constraint that the farmer only has 400 feet of fencing material. As the feet of fencing material is a measure of length, this means that the perimeter of the fence must be 400 feet, or that you are *Given* that the perimeter P satisfies

$$P = 2l + 2w = 400$$

7. $l = \dfrac{400 - 2w}{2} = 200 - w$
8. $A = lw = (200 - w)w = 200w - w^2$
9. It is clear that w can get as close to 0 as you like (with l then getting close to 200), or that w can also get as close to 200 as you like (with l then getting close to 0). Thus the proper interval for w is the open interval $(0, 200)$. You cannot really have w equal to 0 or to 200, or the rectangle will collapse into a line segment. For convenience, however, it's OK to throw in the endpoints, so you can use the methods of Section 6.1. Generally, when you do this, the endpoints will give the worst possible answer, not the best possible answer, so the best possible answer over the interval $[0, 200]$ and the best possible answer over the interval $(0, 200)$ will coincide.
10. $\dfrac{dA}{dw} = 200 - 2w$, so that the derivative is never undefined, but equals 0 when $200 = 2w$, that is, when $w = 100$.
11. $A(100) = 200 \cdot 100 - 100^2 = 20{,}000 - 10{,}000 = 10{,}000$ square feet; $A(0) = A(200) = 0$ square feet.
12. The maximum area is 10,000 square feet. But the problem doesn't ask that—it asks for the *dimensions*. The width w is 100, and you can get the length by going back to Step 7, to see that $l = 200 - w = 200 - 100 = 100$ also. Thus the field should be a 100-by-100-foot square.

Example 7.7 A food processing company plans to introduce a new line of pickles. They plan to sell them in a cylindrical glass jar with a metal top. The glass costs 2 cents per square inch, while the metal top costs 3 cents per square inch. What dimensions will allow the company to produce the most economical jar, if the pickle jar is to have a volume of 200 cubic inches?

Solution:

1. Read the *whole* problem a couple of times.
2. The cylindrical jar can have a small radius and a big height, or radius and height with values close to each other, or a big radius and a small height (see Figure 7.10).
   Of all these configurations, which one will cost the company the least to manufacture?

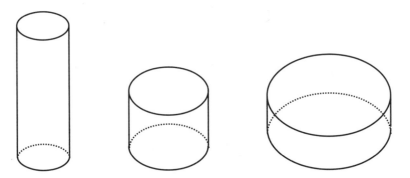

Figure 7.10. *Some of the infinite possibilities*

3. A cylinder is determined most basically by its base radius and its height, so your picture should look like that in Figure 7.11.

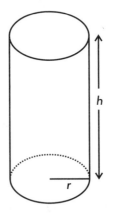

Figure 7.11

4. Even though the problem has the word "most," as part of the phrase "most economical," it is clear that "most economical" means you want to minimize cost.

5. The general formula for cost C is determined by multiplying area by the cost per unit of area. Thus the glass bottom and lateral surface cost 2 cents per square inch, so the cost of the bottom and lateral surface is found by multiplying 2 by the area πr^2 (bottom) $+2\pi rh$ (lateral). The top costs 3 cents per square inch, so the cost of the top is found by multiplying 3 by the area πr^2. Thus the total cost is:

$$C = 2(\pi r^2 + 2\pi rh) + 3\pi r^2 = 5\pi r^2 + 4\pi rh$$

6. The formula for what you want to minimize has two variables, r and h. The constraint or relationship comes from the fact that the volume $V = \pi r^2 h$ is fixed at 200 cubic inches. Thus you are *Given*

$$200 = \pi r^2 h$$

7. If you solve for r in terms of h, you will have to use square roots. Thus it seems easier to try to solve for h in terms of r.

$$h = \frac{200}{\pi r^2}$$

8. $C = 5\pi r^2 + 4\pi r \left(\dfrac{200}{\pi r^2} \right) = 5\pi r^2 + \dfrac{800}{r}$

9. Certainly $r > 0$. There is no upper bound on r. The radius can be as big as you want, and all you must do is make the height h sufficiently small to satisfy the contraint $200 = \pi r^2 h$. Thus r is in the interval $(0, \infty)$.

10. Taking the derivative of the cost formula gives

$$\frac{dC}{dr} = 10\pi r - \frac{800}{r^2}$$

This derivative is undefined for $r = 0$ (which is outside the interval meaningful for the problem), and equals 0 when $10\pi r = \dfrac{800}{r^2}$, that is, for $r^3 = \dfrac{800}{10\pi} = 25.46$, or $r = 2.94$ (approximately).

11. To be sure $r = 2.94$ minimizes cost, use the methods of Section 7.3. The one point $r = 2.94$ on $(0, \infty)$ breaks the interval up into two pieces, on each of which C is either always increasing or always decreasing. Evaluating $\dfrac{dC}{dr} = 10\pi r - \dfrac{800}{r^2}$ at the sample point $r = 1$ in the first interval shows that $\dfrac{dC}{dr} = 31.4 - \dfrac{800}{1} = -768.6$, which shows that C is a decreasing function in the interval $(0, 2.94)$. Evaluating

$\dfrac{dC}{dr} = 10\pi r - \dfrac{800}{r^2}$ at the sample point $r = 4$ in the second interval shows that $\dfrac{dC}{dr} = 40\pi - \dfrac{800}{16} = 125.6 - 50 = 75.6$, which shows that C is increasing on the second interval $(2.94, \infty)$. Thus the point $r = 2.94$ is really where C is the smallest.

12. By now you have done all of the hard work. Don't lose credit by not answering the exact question. The problem asked for the dimensions of the jar. You've found r, but now you must find h. The relationship between r and h was written down in Step 7 as $h = \dfrac{200}{\pi r^2}$. Thus when $r = 2.94$,

$$h = \dfrac{200}{\pi 2.94^2} = \dfrac{200}{27.14} = 7.37 \text{ inches.}$$

Example 7.8 Find the area of the biggest rectangle that can be inscribed in a circle of radius r.

Solution:

1. Read the *whole* problem a couple of times.
2. You are given a circle with a fixed radius. It is called r instead of being given a particular value, so that once you solve the above problem, you will have the answer all at once for all circles. Remember, though, you are not allowed to change the radius of the circle. That is given. Thus, it should be thought of as a constant when you will be taking derivatives. The only variables you have control over are the length and width of the rectangle. A bigger length means a smaller width, and conversely. Figure 7.12 shows some different rectangles that can be inscribed in the circle.

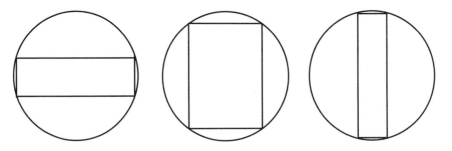

Figure 7.12. *Some of the infinite possibilities*

3. This is clearly a geometry problem. The picture is shown in Figure 7.13.

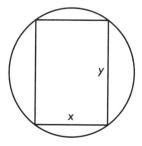

Figure 7.13

4. You want to maximize the area (that's what the phrase "biggest rectangle" tells you).
5. $A = xy$
6. The area A, the quantity to be maximized, is given as a function of two variables. The x and y are related because the rectangle is constrained to fit inside the circle. The relationship may not be immediately apparent from the preceding picture, but this is an example of the extra line trick (see Figure 7.7). From Figure 7.14, you can see that you are *Given* $x^2 + y^2 = (2r)^2 = 4r^2$.

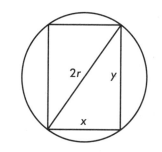

Figure 7.14

7. $y = \sqrt{4r^2 - x^2}$
8. $A = x\sqrt{4r^2 - x^2} = x(4r^2 - x^2)^{\frac{1}{2}}$
9. Clearly $x > 0$ and $x < 2r$. Although x cannot equal either 0 or $2r$, you can for convenience put in the endpoints so you can use the methods of Section 6.1. Thus the interval is $[0, 2r]$.

10. Compute A' first by using the product rule, and then by using the chain rule. Remember, you should treat r like a constant.

$$A'(x) = x \cdot \frac{1}{2}(4r^2 - x^2)^{-\frac{1}{2}} \cdot (-2x) + 1 \cdot (4r^2 - x^2)^{\frac{1}{2}}$$

$$= \frac{-x^2}{(4r^2 - x^2)^{\frac{1}{2}}} + (4r^2 - x^2)^{\frac{1}{2}}$$

$$= \frac{-x^2 + (4r^2 - x^2)^{\frac{1}{2}} \cdot (4r^2 - x^2)^{\frac{1}{2}}}{(4r^2 - x^2)^{\frac{1}{2}}}$$

$$= \frac{-x^2 + (4r^2 - x^2)}{(4r^2 - x^2)^{\frac{1}{2}}}$$

$$= \frac{4r^2 - 2x^2}{(4r^2 - x^2)^{\frac{1}{2}}}$$

Now the derivative is undefined when the denominator $(4r^2 - x^2)^{\frac{1}{2}} = 0$, which happens when $x^2 = 4r^2$, or $x = 2r$, which is an endpoint of the interval. The derivative equals 0 when the numerator equals 0, which happens when $4r^2 - 2x^2 = 0$, or $x = \sqrt{2}r$.

11. From Step 8, $A(0) = 0$, and $A(2r) = 0$. Now

$$A(\sqrt{2}r) = \sqrt{2}r \cdot (4r^2 - 2r^2)^{\frac{1}{2}} = \sqrt{2}r \cdot (2r^2)^{\frac{1}{2}} = \sqrt{2}r \cdot \sqrt{2}r = 2r^2$$

So, the area of the biggest rectangle is $2r^2$. Note from Step 7 that when $x = \sqrt{2}r$ (so that $x^2 = 2r^2$), then also

$$y = \sqrt{4r^2 - x^2} = \sqrt{4r^2 - 2r^2} = \sqrt{2r^2} = \sqrt{2}r$$

Thus the inscribed rectangle of maximum area is in fact a square.

In the previous four examples, the desired extreme value never occurred at an endpoint. To see how it might, suppose Example 7.5 were reworded, so that you wanted to find the smallest possible product you can form from two numbers that are each greater than or equal to 0, and whose sum equals 25. You would do all the work as in the example, but now the minimum product would be 0, which occurs when $x = 0$ or when $x = 25$.

7.3 RELATED RATE WORD PROBLEMS

Implicit Derivatives

Imagine an expanding circle, perhaps formed by throwing a coin into a pool of water. The area A and radius r of the circle of course satisfy $A = \pi r^2$.

However, A and r are also changing with respect to time t, as the circle expands, so that you could consider both A and r as functions of time t. It makes sense to think that the rate at which the area changes with respect to time, $\dfrac{dA}{dt}$, should be related to the rate at which the radius changes with respect to time, $\dfrac{dr}{dt}$, because the area A and radius r are related.

How can you find the relationship between $\dfrac{dA}{dt}$ and $\dfrac{dr}{dt}$? Well, the geometry formula $A = \pi r^2$ gives A as a function of r, and you can *think of r* as a function of t (even if you do not know the formula for how r changes in time). By composing these two functions, you could think of A as a function of t. Recall from Section 5.3 that the *chain rule* tells you how to find the derivative of a composite function.

Let's first review one of the standard versions of the chain rule. Typically, you are given y as a function of u and u as a function of x, so that you can think of y as a function of x also. The chain rule then says that

$$\frac{dy}{dx} = \frac{dy}{du}\frac{du}{dx}$$

Remember, even though the derivatives are *not* fractions, the beauty of the notation is that you can be sure you have the equation written correctly if it just looks correct—that is, if you *thought of* the derivatives as fractions, then in looking at $\dfrac{dy}{du} \cdot \dfrac{du}{dx}$, it *looks* as though du would cancel from the numerator and denominator to give $\dfrac{dy}{dx}$.

One difference in the problem at hand is that the letters are different. Instead of y as a function of u and u as a function of x, you have A as a function of r and r as a function of t. Applying the chain rule with these different letters gives

$$\frac{dA}{dt} = \frac{dA}{dr}\frac{dr}{dt}$$

A second difference in the problem at hand is that while in Section 5.3 you were always given an explicit formula for u in terms of x, here you do not have a formula for r in terms of t. In such a case, the only thing you can do with $\dfrac{dr}{dt}$ is simply to repeat the symbol. As $A = \pi r^2$, clearly $\dfrac{dA}{dr} = 2\pi r$. Thus the chain rule gives

$$\frac{dA}{dt} = \frac{dA}{dr}\frac{dr}{dt} = 2\pi r\frac{dr}{dt}$$

This is how $\dfrac{dA}{dt}$ and $\dfrac{dr}{dt}$ are related.

It is often useful to approach the problem of finding the relationship between $\dfrac{dA}{dt}$ and $\dfrac{dr}{dt}$ a little differently. In this approach, you suppress the letter A, and think of finding the rate of change, with respect to time, of the quantity

πr^2, where r is some function of t. As you do not know a formula for r explicitly in terms of t, you certainly do not have a formula for πr^2 explicitly in terms of t. But you can *think of* πr^2 as depending upon time t, because r does. A derivative with respect to the variable t, like $\dfrac{d(\pi r^2)}{dt}$, is called an **implicit derivative**, because the function you are taking the derivative of, πr^2, is not given *explicitly* in terms of t, but only *implicitly*. That is, you are only *thinking* of it as a function of the time variable t. To find an implicit derivative, you just use the chain rule again. The quantity πr^2 depends upon r and r depends upon t, so that by the chain rule

$$\frac{d(\pi r^2)}{dt} = \frac{d(\pi r^2)}{dr}\frac{dr}{dt} = 2\pi r\frac{dr}{dt}$$

To find the relationship between $\dfrac{dA}{dt}$ and $\dfrac{dr}{dt}$ in this approach, you look at the relationship between A and r, namely, $A = \pi r^2$, and think of taking the implicit derivative of both sides with respect to the variable t. With the symbol A, there's not much you can do. The derivative of A with respect to t is just $\dfrac{dA}{dt}$. For the implicit derivative of the right-hand side, πr^2, with respect to t, you just do as before. Thus,

$$
\begin{aligned}
A &= \pi r^2, \text{ so that}\\
\frac{d}{dt}(A) &= \frac{d}{dt}(\pi r^2)\\
\frac{dA}{dt} &= \frac{d(\pi r^2)}{dt}\\
\frac{dA}{dt} &= \frac{d(\pi r^2)}{dr}\frac{dr}{dt}\\
\frac{dA}{dt} &= 2\pi r\frac{dr}{dt}
\end{aligned}
$$

With a little practice you won't have to write out all these steps. You should become able to do a number of them at once in your head.

Example 7.9 Suppose that x and y are functions of time t. Find the rate of change, with respect to time, of the expression $x^2 y^3$.

Solution: You want to compute the implicit derivative

$$\frac{d(x^2 y^3)}{dt}$$

Here you are thinking of x and y as both being functions of time t. First of all, the expression $x^2 y^3$ is a *product* of x^2 and y^3, so you use the product rule to get

$$\frac{d(x^2 y^3)}{dt} = x^2\frac{d(y^3)}{dt} + \frac{d(x^2)}{dt}y^3 \tag{7.1}$$

To compute the implicit derivative $\dfrac{d(y^3)}{dt}$, use the chain rule:

$$\frac{d(y^3)}{dt} = \frac{d(y^3)}{dy}\frac{dy}{dt} = 3y^2\frac{dy}{dt}$$

Likewise, to compute the implicit derivative $\frac{d(x^2)}{dt}$, use the chain rule:

$$\frac{d(x^2)}{dt} = \frac{d(x^2)}{dx}\frac{dx}{dt} = 2x\frac{dx}{dt}$$

Substituting these two derivatives into Equation 7.1 gives

$$
\begin{aligned}
\frac{d(x^2 y^3)}{dt} &= x^2\frac{d(y^3)}{dt} + \frac{d(x^2)}{dt}y^3 \\
&= x^2 3y^2\frac{dy}{dt} + 2x\frac{dx}{dt}y^3 \\
&= 3x^2 y^2\frac{dy}{dt} + 2xy^3\frac{dx}{dt}
\end{aligned}
$$

The Word Problems

All related rate word problems give you information concerning the rates of change of certain quantities, and ask you to find the rate of change of some other quantity. To do this, you must find a relationship between the quantities whose rate of change is given and the quantity whose rate of change you are to find. Often the problem is geometric in nature, and the relationship can be found once you draw the picture. After finding the relationship, take the derivative with respect to time of both sides. This will often involve implicit derivatives. Then substitute all the information the problem gives you about values at a particular instant, and solve. It is critical that information about values at a particular instant not be substituted until *after* you have taken the derivative.

Below is an enumeration of the steps to go through. Again, reading comprehension is the most important skill you need to do word problems successfully. After the enumeration, you will find some worked examples.

1. Read the problem once over quickly.
2. Read the problem over carefully, and be sure you can understand it. Try to rephrase it in your own words.
3. If the problem involves geometry, draw a picture and label the picture. You can label as constant anything that stays constant throughout the course of the problem. If a quantity varies throughout the course of the problem, you must label it with a variable (i.e., letter), even if the problem ultimately asks you for information when that quantity has a particular value, such as 60.

4. Write down which derivatives you are given. Look for key words such as "velocity" or "speed" (the derivatives of position with respect to time), or the word "rate" (such as "the balloon is being filled at the rate of 45 cubic feet per second").

5. Write down the derivative you are asked to find.

6. Look at the quantities whose derivatives you are given and the quantity whose derivative you are asked to find, and find a relationship between all of these quantities.

7. Viewing all of the above quantities as functions of time, take an implicit derivative with respect to time of both sides of the relationship.

8. Substitute any particular information the problem gives you about values of quantities at a particular instant, and solve the problem. To find all of the values to substitute, you may have to use once more the relationship you found in Step 6.

Example 7.10 Two cars leave an intersection at 1 P.M. The first car is traveling east at 45 miles per hour, while the second car is traveling north at 60 miles per hour. How fast is the distance between the cars increasing at 3 P.M.

Solution:

1. Read the *whole* problem a couple of times.

2. The distance must mean the straight-line distance. The speeds of the cars are their rates of change of position with respect to time.

3. The relevant picture is the right triangle (see Figure 7.15).

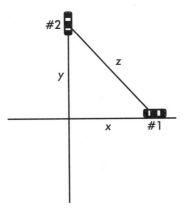

Figure 7.15

4. You are given the speed of the first car, $\dfrac{dx}{dt} = 45$. You are also given the speed of the second car, $\dfrac{dy}{dt} = 60$.

5. You are asked to find how fast the distance between the cars is increasing, that is, the rate of change of distance between the cars with respect to time. If you label the distance between the cars as z, you are asked to find $\dfrac{dz}{dt}$.

6. You now want to find a relationship between x, y, and z. It is clear from the Pythagorean theorem that the relationship is

$$x^2 + y^2 = z^2$$

7. Take an implicit derivative with respect to time of both sides. Remember that x, y, and z are all changing in time. If you have to, write out the full version of the chain rule, that is, for example,

$$\frac{d(z^2)}{dt} = \frac{d(z^2)}{dz}\frac{dz}{dt} = 2z\frac{dz}{dt}$$

Hopefully, though, you can see the pattern of what is going on, and just do the middle step in your head, and simply write

$$\frac{d(z^2)}{dt} = 2z\frac{dz}{dt}$$

That's what I will do!

$$\frac{d(x^2 + y^2)}{dt} = \frac{d(z^2)}{dt}$$

$$2x\frac{dx}{dt} + 2y\frac{dy}{dt} = 2z\frac{dz}{dt}$$

8. You have six quantities in the previous equation, and you want to solve for one of them, $\dfrac{dz}{dt}$. This means you should be able to substitute values for the other five of them, and then solve. You have already recorded that $\dfrac{dx}{dt} = 45$ and that $\dfrac{dy}{dt} = 60$. What about x, y, and z? This is where reading comprehension comes in. If the first car is traveling east at 45 miles per hour, and it has traveled for two hours (the time span between 1 P.M. and 3 P.M.), then it has traveled $45 \cdot 2 = 90$ miles, so that $x = 90$. Likewise, $y = 60 \cdot 2 = 120$ miles. What about z? Here you have to use the relationship you found in Step 6.

$$
\begin{aligned}
x^2 + y^2 &= z^2 \\
90^2 + 120^2 &= z^2 \\
8{,}100 + 14{,}400 &= z^2 \\
\sqrt{22{,}500} = 150 &= z
\end{aligned}
$$

Substituting all of this gives

$$2 \cdot 90 \cdot 45 + 2 \cdot 120 \cdot 60 = 2 \cdot 150 \cdot \frac{dz}{dt}$$

Thus,

$$\frac{dz}{dt} = \frac{8{,}100 + 14{,}400}{300} = \frac{22{,}500}{300} = 75 \text{ miles per hour}$$

Again, note that it would have been a mistake to substitute 90 for x or 120 for y *before* you took the derivative. Note also that in this example it would have been possible to get explicit formulas for x and y, and thus for z also, in terms of t. However, the above method is generally simpler.

Example 7.11 Air is being pumped into a spherical balloon at the rate of 20 cubic feet per second. At what rate is the radius of the balloon increasing at the instant when the radius equals 15 feet?

Solution:

1. Read the *whole* problem a couple of times.
2. As air is being pumped in, both the radius and the volume get bigger. The problem asks about the rate of change of radius with respect to time.
3. The picture isn't too much help, but here it is (see Figure 7.16).

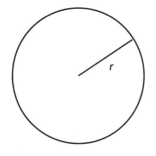

Figure 7.16

4. Note that cubic feet is a unit of volume. Thus as air is being pumped in at the rate of 20 cubic feet per second, you have as a given that $\dfrac{dV}{dt} = 20$.

5. You want to find $\dfrac{dr}{dt}$.

6. Looking at the previous two steps, you see that you want to find a relationship between V and r. To do this, you must already know the formula for the volume of a sphere as a function of its radius, $V = \dfrac{4}{3}\pi r^3$.

7. Take the derivative with respect to time of both sides. Remember, V and r are both changing in time.

$$\frac{dV}{dt} = \frac{d\left(\frac{4}{3}\pi r^3\right)}{dt}$$

$$\frac{dV}{dt} = \frac{4}{3}\pi \cdot 3r^2\frac{dr}{dt}$$

$$= 4\pi r^2\frac{dr}{dt}$$

Note that as the left-hand side of the relationship was just V, there is nothing to write for the derivative of V with respect to t except the symbol for that derivative, $\dfrac{dV}{dt}$. For the derivative with respect to t of the right-hand side of the relationship, you must think of r as a function of t. Then $\dfrac{d\left(\frac{4}{3}\pi r^3\right)}{dt} = \dfrac{4}{3}\pi\dfrac{d(r^3)}{dt}$, and for $\dfrac{d(r^3)}{dt}$ you use the chain rule.

8. There are three quantities in the derivative formula of Step 7, namely, $\dfrac{dV}{dt}$, r, and $\dfrac{dr}{dt}$. As you want to find $\dfrac{dr}{dt}$, you must substitute for the other two. You have already seen that $\dfrac{dV}{dt} = 20$. Note that the problem states that you want to find the rate at which the radius is increasing at the instant when the radius is 15. Now that you have already taken the derivative, you can substitute the particular value of the radius at that instant, $r = 15$. Thus,

$$20 = 4\pi(15)^2\frac{dr}{dt}$$

$$20 = 900\pi\frac{dr}{dt}$$

$$\frac{20}{900\pi} = \frac{1}{45\pi} = \frac{dr}{dt}$$

The radius of the balloon is increasing at the rate of $\dfrac{1}{45\pi} > 0.0071$ feet per second.

Example 7.12 A truck is dumping sand into a conical pile at the rate of 30 cubic feet per second, and in such a way that the height of the pile is always equal to twice the radius. At what rate is the height increasing when the sand pile has a volume of 300 cubic feet?

Solution:

1. Read the *whole* problem a couple of times.
2. As the sand is being dumped on, the volume of the pile and its radius and its height are all changing.
3. The cone is of course a cone with its point up in the air (see Figure 7.17).

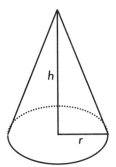

Figure 7.17

4. You are given that $\dfrac{dV}{dt} = 30$. To see this, remember that cubic feet is a measure of volume.
5. You want to find $\dfrac{dh}{dt}$.
6. You thus need to find a relationship between V and h. Now the standard formula for the volume of a cone is $V = \dfrac{1}{3}\pi r^2 h$. As the radius r is also changing in this problem, to get a relationship between V and h, you must express r in terms of h. The words in the problem statement "the height of the pile is always equal to twice the radius" tell you how to do this. The words state that $h = 2r$, so that $r = \dfrac{1}{2}h$. Thus,

$$
\begin{aligned}
V &= \frac{1}{3}\pi r^2 h \\
&= \frac{1}{3}\pi \left(\frac{h}{2}\right)^2 h \\
&= \frac{\pi}{12}h^3
\end{aligned}
$$

7. It follows that

$$\frac{dV}{dt} = \frac{d\left(\frac{\pi}{12}h^3\right)}{dt}$$

$$= \frac{\pi}{12}3h^2\frac{dh}{dt}$$

$$= \frac{\pi}{4}h^2\frac{dh}{dt}$$

8. First substitute into the previous equation the value of $\frac{dV}{dt} = 30$. You next have to substitute in a value for h. How can you do this? Well, you know that at the instant when the problem asks for $\frac{dh}{dt}$, you have $V = 300$. You have also worked out in Step 6 a relationship between V and h, namely, $V = \frac{\pi}{12}h^3$. Substitute $V = 300$ in this relationship to find h. Thus,

$$300 = \frac{\pi}{12}h^3$$

$$\frac{300 \cdot 12}{\pi} = h^3$$

$$1146.5 = h^3$$

$$10.47 = h$$

Substituting $\frac{dV}{dt} = 30$ and $h = 10.47$ into the derivative formula of Step 7 gives

$$30 = \frac{\pi}{4}(10.47)^2\frac{dh}{dt}$$

$$\frac{120}{109.62\pi} = \frac{dh}{dt}$$

$$0.35 = \frac{dh}{dt}$$

The height is increasing at a rate of 0.35 feet per second.

7.4 SUMMARY OF MAIN POINTS

- Word problems are first and foremost exercises in reading comprehension. You must understand the English, and translate it into mathematical symbols. To do this, you must also be able to explain in English what the mathematical symbols mean.

- Read each word problem over once quickly, to get the general idea, and then more carefully. Be sure you really understand what the problem asks for. Can you restate the problem in your own words?

- Most word problems involve geometry. For these problems, you should draw a picture and label the parts of the picture. You might have to use some standard geometry formulas to work the problem.

- In a problem involving geometry, always be aware of the difference between length, area, and volume. Also be aware of the possibility that you might have to draw an extra line into the picture, at just the right spot, to see the needed relationships.

- For max-min problems, be sure you know, and write down, whether you want to maximize or minimize, and which quantity you want to maximize or minimize. To help identify the quantity, look for superlative adjectives or adverbs.

- Related rate word problems most often require that you compute an *implicit* derivative. To do this, you must understand the chain rule well. Thus, if y is a quantity that changes throughout the course of a problem, and you are computing the derivative with respect to time of y^2, you must use the chain rule and compute

$$\frac{d(y^2)}{dt} = \frac{d(y^2)}{dy}\frac{dy}{dt} = 2y\frac{dy}{dt}$$

- For related rate word problems, be sure you identify, and write down as *Given*, those quantities whose derivatives you are given. Also identify, and write down as *Find*, the quantity whose derivative you want to find. To help in identifying these quantities, look for the word "rate," or phrases such as "how fast is...?" Also remember that speed is the rate of change of position with respect to time.

- For the rest of the specific steps to guide you in solving maximum-minimum word problems, see Section 7.2.

- For the rest of the specific steps to guide you in solving related rate word problems, see Section 7.3.

7.5 EXERCISES

Solve the max-min word problems in Exercises 1–13.

1. A farmer has 400 feet of fencing and wants to enclose a rectangular pasture. One side of the pasture is along a river, and does not need fencing. What should the dimensions of the rectangular pasture be, so as to maximize the enclosed area?

2. A landscape architect wants to fence in a rectangular garden area of 2,500 square feet. What dimensions will require the least amount of fencing material?

3. A rectangular box with a square base is to be made from two different materials. The material for the top and four sides costs $1 per square foot, while the material for the bottom costs $2 per square foot. If you can spend $196 on materials, what dimensions will maximize the volume of the box?

4. A nursery school wants to fence in a rectangular playground. One side of the playground is along a busy street, and needs taller fencing that costs $10 per foot. The fencing for the other three sides costs $7 per foot. If the school has a budget of $6,000, what are the dimensions of the playground of largest area that they can build?

5. Find the area of the largest rectangle that can be inscribed in a *semi*circle of radius 4 inches.

6. Find the area of the largest rectangle that can be inscribed in a semicircle of radius a inches.

7. A right circular cylinder is to have a volume of 300 cubic inches. What dimensions will minimize the *total* surface area of the cylinder?

8. A right circular cylinder is to have volume of 300 cubic inches. Assuming that the cylinder has no top, what dimensions will minimize the remaining surface area of the cylinder?

9. The material for the bottom of an open-topped cylindrical pot costs 4 cents per square inch, while the material for the curved surface costs 2 cents per square inch. If the costs of materials are to be held to $10 per pot, what is the maximum volume pot that can be manufactured?

10. Find the two nonnegative numbers x and y whose sum is 50, and for which the product $P = x^2y$ is maximized.

11. Find the two nonnegative numbers x and y whose sum is 50, and for which the product $P = xy$ is minimized.

12. Find the two nonnegative numbers whose product is 100, and whose sum is minimized.

13. Find the two nonnegative numbers whose product is 100 and whose sum is maximized.

Solve the related rate problems in Exercises 14–28.

14. Car A passes through an intersection, traveling east, at 10 A.M., and maintains a constant velocity of 60 mi/h. Car B travels north through the same intersection at 11 A.M., and maintains a constant velocity of 50 mi/h. How fast is the distance between the cars increasing at noon?

15. The top of a fifteen-foot ladder is sliding down a wall, while the foot of the ladder is sliding along the floor. If the top of the ladder is sliding

down the wall at the rate of 2 inches per second, with what speed is the foot of the ladder sliding along the floor, when the foot of the ladder is 5 feet from the wall?

16. A girl is flying a kite, and is of course holding on to the kite with a kite string. If the kite is 100 feet up in the air, and is being blown horizontally by the wind at a speed of 40 feet per second away from the girl, at what rate is the length of kite string that is out increasing, when 125 feet of string are already out?

17. Juan is standing 10 feet away from Eleni when she releases a helium-filled balloon that flies straight up into the air at a speed of 20 feet per second. At what rate is the distance between Juan and the balloon increasing, when the balloon is 30 feet in the air?

18. With the same information as in Exercise 17, at what rate is the distance between Juan and the balloon increasing when the balloon is 30 feet from Juan?

19. A truck is dumping sand into a conical pile at the rate of 60 cubic feet per second, and in such a way that the height of the pile is always equal to three times the radius of the base. At what rate is the height of the sand pile increasing at the instant when the height is 20 feet?

20. With the same information as in Exercise 19, at what rate is the height of the sand pile increasing when the radius of the base of the pile is 20 feet?

21. A conical paper cup has a height of 6 inches, while the radius of the circular opening at the top is 2 inches. The cup was initially filled with water, but there is a leak at the bottom. At what rate is water leaking out of the cup when the water level in the cup is 3 inches, and is dropping at the rate of 1 inch per minute?

22. With the same information as in Exercise 21, at what rate is water leaking out of the cup when the water level is dropping at the rate of $\frac{1}{2}$ inch per minute, and there are exactly 16 cubic inches of water left in the cup?

23. A spherical hot air balloon is being filled with air at the rate of 200 cubic feet per minute. At what rate is the radius of the balloon increasing when the balloon has 1000 cubic feet of air in it?

24. A spherical balloon is being filled with air at a constant rate. It is observed that when the radius of the balloon is 14 feet, the radius is increasing at the rate of 6 feet per minute. What is the rate at which air is being pumped into the balloon?

25. A stone thrown into a pond creates a circular ripple. If the radius of the circle is increasing at the rate of 2 feet per second, at what rate is the area of the circle increasing when the radius is 12 feet?

26. With the same information as in Exercise 25, at what rate is the area of the circle increasing when the area is 100 square feet?

27. A particle is traveling around the circle $x^2 + y^2 = 25$ in such a way that the rate of change of its x-coordinate is $\dfrac{dx}{dt} = 2$. Find the rate of change $\dfrac{dy}{dt}$ of its y-coordinate when the particle is at the point $(3, -4)$ on the circle.

28. A particle is sliding down the curve $y = \dfrac{1}{x}$. When it is at the point $(1, 1)$ on the curve, $\dfrac{dy}{dt} = -2$. Find $\dfrac{dx}{dt}$ at this instant.

8
THE IDEA OF THE INTEGRAL

Calculus is the major mathematical toolkit for studying and describing those quantities in the world around us that are continuously changing. Of all the ideas of calculus, the most fundamental and useful is that of the **integral**, which will be explained in this chapter. Techniques for computing some integrals from the definition will be discussed in Chapter 9, while Chapter 10 provides an easier alternative for computing integrals.

8.1 THE BASIC IDEA

There are a number of problems that are easy to solve, or computations that are easy to perform, as long as certain parts of the problem situation or computational situation remain constant. By means of the integral, such problems can also be solved when these parts of the problem are changing or variable.

Following are two examples of problems that are easy to solve, because the important quantities in the problems remain constant.

Example 8.1 A bowling ball is rolling down the lane with a constant velocity of 9 feet per second. How far does it travel during the first two seconds of its roll?

> **Solution:** The fact that the bowling ball has a constant velocity of 9 feet per second simply means that the ball rolls 9 feet for every second that it is rolling. Thus in 2 seconds it rolls $9 \cdot 2 = 18$ feet. The distance the ball rolls is just its change in position. As velocity is the rate of change of position with respect to time, then with constant velocity,
>
> $$\text{velocity} = \frac{\text{change in position}}{\text{change in time}}$$
>
> Thus,
>
> $$\text{change in position} = \text{velocity} \cdot \text{change in time} \qquad (8.1)$$

Example 8.2 Find the area of the region in the xy-plane bounded below by the x-axis, on the left by the line $x = 1$, on the right by the line $x = 5$, and above by the line $y = 3$.

Solution: The region is pictured in Figure 8.1.

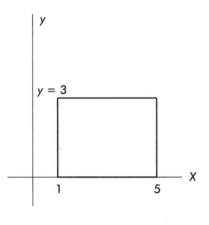

Figure 8.1

Notice that the region is just a rectangle. The area of a rectangle is simply its height times its width. The height of the rectangle is 3 (the distance between the line $y = 3$ and the x-axis $y = 0$ is $3 - 0 = 3$), while the width of the rectangle is 4 (the distance between the line $x = 1$ and the line $x = 5$ is $5 - 1 = 4$). Thus the area of the rectangle is $3 \cdot 4 = 12$. For a rectangle,

$$\text{area} = \text{height} \cdot \text{width}$$

Roughly speaking, when a problem can be solved as long as the "problem ingredients" are constant (and especially when there is a multiplication involved in the solution, as in the above two examples), then by using ideas of this chapter and the following two chapters, the problem can be solved when one or more of the "problem ingredients" is changing or is variable. Following are two examples of problems of this type, related to Examples 8.1 and 8.2. You will revisit these two problems throughout the next several chapters. In this chapter, you will learn how to *estimate* a solution to these problems. By the end of Chapter 9 you will be able to solve them *exactly*, and by the end of Chapter 10 you will be able to solve them both exactly and *easily*.

Problem 1 A bowling ball is thrown down the lane with an initial velocity of 9 feet per second, but due to friction with the floor, it is continually slowing down, so that its velocity t seconds after it is thrown is $v(t) = 9 - 0.2t$. How far does it travel during the first 2 seconds of its roll?

Problem 2   Find the area of the region in the xy-plane bounded below by the x-axis, on the left by the line $x = 1$, on the right by the line $x = 5$, and above by the curve $y = x^2$.

The region is pictured in Figure 8.2.

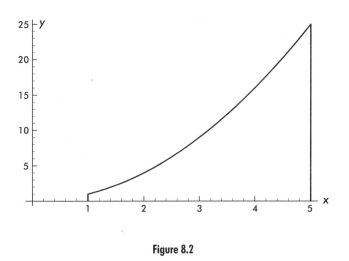

Figure 8.2

You can see that in each of Problems 1 and 2, one of the constant quantities from Examples 8.1 and 8.2 has been made into a variable, changing quantity, so that the problem is no longer able to be solved simply by a multiplication. The constant velocity of 9 feet per second in Example 8.1 was replaced in Problem 1 by the velocity function $v(t) = 9 - 0.2t$, while the constant height of the top $y = 3$ for the region of Example 8.2 was replaced by the changing height of the top $y = x^2$ for the region of Problem 2.

The key *ideas* for solving Problems 1 and 2 are simple and beautiful, although the *calculations* needed to get an exact answer can get complicated. The first idea involves how to *estimate* the correct answer. The second idea involves how to get better and better estimates.

Estimating an Answer for Problem 1

The key idea for Problem 1 is to notice that the velocity of the bowling ball does not change very much over a very small time interval. For example, at time $t = 1$ the velocity is $v(1) = 9 - 0.2 \cdot (1) = 8.8$ feet per second, while at time $t = 1.2$, the velocity is $v(1.2) = 9 - 0.2 \cdot (1.2) = 9 - 0.24 = 8.76$ feet per second. Thus over the time interval between 1 and 1.2 seconds after the ball is released, the velocity doesn't change by very much, and so it is approximately constant (that is, practically constant).

What number should you use as a sample or representative value for the velocity of the ball in the short time interval between $t = 1$ and $t = 1.2$? Well, the velocity of the ball at the beginning $t = 1$ of the time interval is $v(1) = 9 - 0.2 \cdot (1) = 8.8$ feet per second, while the velocity of the ball at the end $t = 1.2$ of the time interval is $v(2) = 9 - 0.2 \cdot (1.2) = 8.76$ feet per second. Either of these two numbers, or their average, or the velocity at the middle $t = 1.1$ of the time interval, or indeed any number between 8.8 and 8.76 would be fine. The idea is that if you are assuming that the velocity of the ball during the short time interval from $t = 1$ to $t = 1.2$ does not change by very much, then you could take the velocity at *any* instant of time in that interval as a good sample or representative value of velocity.

Notice also that the *duration* of the short time interval is determined by the *change* in time from $t = 1$ to $t = 1.2$, and that change is computed by subtraction, so that the change in time is $1.2 - 1 = 0.2 = \dfrac{2}{10} = \dfrac{1}{5}$.

Using the midpoint $t = 1.1$, and the velocity $v(1.1) = 9 - 0.2 \cdot (1.1) = 9 - 0.22 = 8.78$ feet per second as a representative velocity, you can estimate the distance traveled over the small time interval from $t = 1$ to $t = 1.2$:

approximate distance traveled

= approximately constant velocity · change in time

Thus,

approximate distance traveled $= 8.78 \cdot (1.2 - 1) = 8.78 \cdot (0.2) = 1.756$ feet

What about the total distance traveled by the ball for the first 2 seconds of its roll (that is, the distance it travels between times $t = 0$ and $t = 2$)? Well, what you did above for the fifth-of-a-second time interval between $t = 1$ and $t = 1.2$ can be repeated for *each* fifth-of-a-second time interval between $t = 0$ and $t = 2$ (for a total of ten fifths of a second).

Over each fifth-of-a-second interval, the velocity of the ball does not change very much, and so is approximately constant. What constant number should you pick, in each fifth-of-a-second time interval, to represent the velocity of the ball during that interval? Well, if you really believe that the velocity of the ball doesn't change much, then *any* value of the velocity function during the small time interval would be fine as a representative value. You just have to pick one. To be consistent with what we did for the time interval from $t = 1$ to $t = 1.2$, let's pick the velocity at the midpoint of each time interval as a representative value.

Multiply the representative value of velocity during each time interval by the duration 0.2 of the time interval that the ball rolls at that representative velocity to get an estimate of the distance the ball travels during each time interval of one-fifth of a second. Clearly the total distance the ball rolls is simply the sum of the distances it rolls in each of the ten fifth-of-a-second time intervals. As you now have estimates for these distances, you can just add them up and get an estimate for the total distance the ball rolls. The table below shows the results.

Time interval	Mid-point of time interval	Velocity at midpoint, using $v(t) = 9 - 0.2t$	Estimated distance ball rolls during time interval: distance = velocity · time
$[0, 0.2]$	0.1	8.98	1.796 ft
$[0.2, 0.4]$	0.3	8.94	1.788 ft
$[0.4, 0.6]$	0.5	8.9	1.78 ft
$[0.6, 0.8]$	0.7	8.86	1.772 ft
$[0.8, 1.0]$	0.9	8.82	1.764 ft
$[1.0, 1.2]$	1.1	8.78	1.756 ft
$[1.2, 1.4]$	1.3	8.74	1.748 ft
$[1.4, 1.6]$	1.5	8.7	1.74 ft
$[1.6, 1.8]$	1.7	8.66	1.732 ft
$[1.8, 2.0]$	1.9	8.62	1.724 ft

The sum of the numbers in the right-hand column equals 17.6 feet, which represents the *estimate* of the distance the ball rolls.

Getting Better and Better Estimates for Problem 1

You can only expect the answer of 17.6 feet for the distance the bowling ball rolls in two seconds to be an approximation of the true distance it rolls. This is because the basic idea of replacing the changing velocity in a very short time interval by a constant velocity leads only to an *estimate* of the distance the ball rolls in that short interval. Although the velocity of the ball in a short time interval is approximately constant, it is not exactly constant. It doesn't vary by much, but it does vary!

If, however, you replace the time intervals of one-fifth of a second by even shorter time intervals, say of one-fiftieth of a second, you could reasonably expect the velocity function to vary less in each time interval, and thus be more approximately constant. Then the estimate for the distance the ball rolls in the shorter time interval would be more accurate, and if you add up these more accurate estimates of distance rolled for all the time intervals, you could expect a more accurate estimate for the total distance the bowling ball rolls in two seconds.

At this stage, you might be thinking, "But that means estimating the distances the ball rolls for each of the *one hundred* short time intervals of duration one-fiftieth of a second between times $t = 0$ and $t = 2$. That would require a table ten times as long as the previous one. How could anyone do all of that work?" Well, just as in Section 2.2, "Instantaneous Rates of Change," cumbersome numerical work can and will be replaced by algebraic work. Not only that, but by using *limits*, you will be able to compute an *exact* answer as the *limit* of a collection of better and better estimates. The algebra and full description for all of this is in the rest of this chapter and in Chapter 9.

Estimating an Answer for Problem 2

Look at Problem 2 again. The problem was to find the area of the region in the xy-plane bounded below by the x-axis, on the left by the line $x = 1$, on the right by the line $x = 5$, and above by the curve $y = x^2$. Obviously, the region is not a rectangle precisely because the top is given by the varying function $y = x^2$, rather than by a constant function, as in Example 8.2. However, look at a small interval on the x-axis, say between $x = 3$ and $x = 3.4$, and sketch the region in the xy-plane bounded below by the x-axis, on the left by the line $x = 3$, on the right by the line $x = 3.4$, and above by the curve $y = x^2$ (see Figure 8.3).

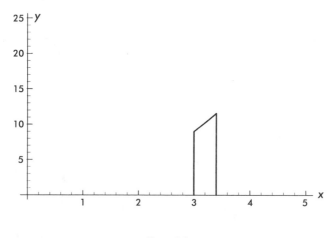

Figure 8.3

This region looks like a thin vertical strip, and the different points on the top do not vary as much in height as they do in the original region pictured in Figure 8.2. This is obviously because the values of the function $y = x^2$ vary less for x in the interval from 3 to 3.4 than they do for x in the interval from 1 to 5. You can thus make the *estimate* that the thin vertical strip is approximately a rectangle. If so, then its area is approximately equal to its width times its height. The width is easy—it's $3.4 - 3 = 0.4$. What value should you use for the height? Well, the height of the left side of the strip is $y(3) = 3^2 = 9$, while the height of the right side of the strip is $y(3.4) = 3.4^2 = 11.56$. Either of these two numbers, or the average, or the height of the strip at the midpoint $x = \dfrac{3 + 3.4}{2} = 3.2$ of the interval, or indeed any number between 9 and 11.56 would be fine. The idea is that if you are making the estimate that the vertical strip is approximately a rectangle, and that the y-coordinates of the points on the top do not vary very much, then you could take the y-coordinate of *any* point on the top as a good representative for the height of the strip.

Using the midpoint $x = 3.2$ and the height $y(3.2) = 3.2^2 = 10.24$ as a representative height, the area of the strip is approximately given by

$$\text{approximate height} \cdot \text{width} = 10.24 \cdot (0.4) = 4.096$$

What about the total area under the graph and over the x-axis, between $x = 1$ and $x = 5$? Well, what you did above for the small interval between $x = 3$ and $x = 3.4$ can be repeated for each of the ten small intervals of width 0.4 between $x = 1$ and $x = 5$. The picture is shown in Figure 8.4.

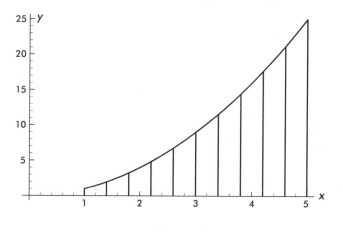

Figure 8.4

Over each little interval the values of $y = x^2$ do not change as much as they do over the whole interval $[1, 5]$. Thus you could *estimate* the area of each little vertical strip by pretending it is a rectangle, with area $=$ height $\cdot$ width. The width of each little strip is 0.4, and for the height of one of the thin vertical strips, you could reasonably use the value of $y = x^2$ for *any* value of x in the little interval.

Multiply the representative value of height in each little interval by the width 0.4 of the interval to get an estimate for the area of each thin vertical strip. Clearly the area of the whole region is simply the sum of the areas of the vertical strips, so you can get an *estimate* for the area of the whole region simply by adding the estimates for the areas of each strip. In the following table, the height of the strip is estimated by computing $y = x^2$ at the midpoint of each interval.

Interval	Midpoint of interval	Height at midpoint, using $y = x^2$	Estimated area of interval: area = height · width (each width = 0.4)
$[1, 1.4]$	1.2	1.44	0.576
$[1.4, 1.8]$	1.6	2.56	1.024
$[1.8, 2.2]$	2.0	4.0	1.6
$[2.2, 2.6]$	2.4	5.76	2.304
$[2.6, 3.0]$	2.8	7.84	3.136
$[3.0, 3.4]$	3.2	10.24	4.096
$[3.4, 3.8]$	3.6	12.96	5.184
$[3.8, 4.2]$	4.0	16.0	6.4
$[4.2, 4.6]$	4.4	19.36	7.744
$[4.6, 5.0]$	4.8	23.04	9.216

The sum of the numbers in the right-hand column equals 41.28, which represents the *estimate* of the area of the region. Pictured in Figure 8.5 are the rectangles whose areas you actually computed. The height of each thin rectangle is given by the value of $y = x^2$ at the midpoint of each thin interval.

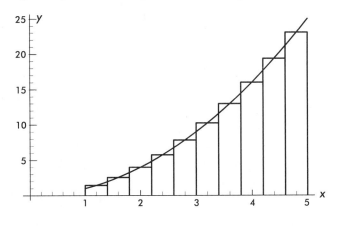

Figure 8.5

Getting Better and Better Estimates for Problem 2

You can only expect the answer of 41.28 for the area of the region to be an approximation of the true area. This is because the idea of replacing each thin strip by a thin rectangle leads only to an *estimate* of the area of the thin strip. Each thin strip is approximately a rectangle, but it is not exactly a rectangle. The heights of the points on top do not vary by too much, but they do vary.

If, however, you replace the intervals of width 0.4 by thinner intervals of width 0.1, you could reasonably expect the heights of the points on top of the thinner intervals to vary by much less, and thus for the area of each thinner vertical strip to be more accurately approximated by the area of a rectangle. Then if you add up the areas of all these rectangles, you should get a much more accurate estimate for the area of the original region. The picture of the vertical strips and the approximating rectangles is shown in Figure 8.6, and you can see that the idea seems to be right.

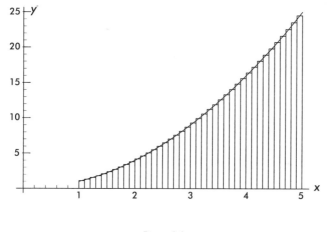

Figure 8.6

Without a computer or a programmable calculator, the computations, of course, would be very tedious and time consuming. But as mentioned when discussing Problem 1, by using algebra and *limits*, you can get the exact answer as a limit of estimates, without doing all that much numerical work.

To get a glimpse of the unifying power of mathematics, compare the above discussion of the area of a region in the plane with the previous example of the distance the bowling ball rolls. One problem talks about time, velocity, and distance, while the other talks about widths of intervals, height, and area. They are each talking about completely different concepts, but the procedures used to arrive at an answer are remarkably similar.

The basic idea, again, is that if you have a problem you can solve when all of the quantities involved are constant, then if the problem becomes difficult when one of the quantities is changing, try breaking the problem up into lots of little pieces. If on each little piece the changing quantity is approximately constant, then you can solve the problem approximately for each little piece. If the solution for the whole problem can be found by adding up the solutions for each little piece of the problem, then you have an approximate solution for the whole problem. To get a better approximation, break the problem up into even smaller pieces. If, as you break the problem up into more and more smaller and smaller pieces, your approximations seem to be getting closer and closer to some one particular number, then you are close to arriving at an exact answer.

8.2 TERMINOLOGY AND NOTATION

In this section, you will find the terminology and notation that can be used to describe more *systematically* and *algebraically* the work done in the previous section. The next chapter, "Computing Some Integrals," will show you how to obtain the exact answers to Problems 1 and 2, the bowling ball and area problems of Section 8.1 (see pages 155 and 156).

As you saw in Chapter 2 on "Change, and the Idea of the Derivative," there is a great advantage to be gained in replacing numbers with letters, in substituting algebra for arithmetic. The advantage is that many different problems (the velocity of the ball at time 2, the velocity of the ball at time 3, the velocity of the ball at time 4) become one problem (the velocity of the ball at time t). The price you pay for this advantage is that something relatively concrete (a numerical computation, with numbers) becomes more abstract (an algebraic computation, with letters and symbols). To still be able to keep a grip on what is going on, you really have to have a good idea of what the symbols mean.

Another advantage of using symbols and technical terminology is that one sentence or one line can replace a short essay. This makes for economical communication, but you've really got to keep the short essay (that is, what the symbols mean) in your mind to know what's going on.

In Problems 1 and 2, you had four ingredients:

1. a variable (t for time in the bowling ball problem, x for position on the x-axis in the area problem);
2. a function of that variable ($v(t) = 9 - 0.2t$ for the velocity of the bowling ball, $y = f(x) = x^2$ for the top of the region);
3. an interval over which you were considering the variable (time t varied from $t = 0$ to $t = 2$ in the bowling ball problem, so the interval there was $[0, 2]$, while position x varied from $x = 1$ to $x = 5$ in the area problem, so the interval for the area problem was $[1, 5]$);
4. a problem to be solved that, if the function in item 2 were constant, would be easy to solve by multiplying the value of the function times the length of the interval.

The most standard notations for the first three ingredients above are

1. x for the variable;
2. $f(x)$ for the function;
3. $[a, b]$ for the interval, so that you are looking at the interval $a \leq x \leq b$.

The terminology and notation can be set up without having a specific problem in mind, as the technique applies to remarkably many different problems. If you want to keep something specific in mind, think about the distance the bowling ball rolls, or the area under the graph.

Notation and Terminology for "Dividing the Interval into Small Pieces"

One idea of the previous section was to break up or partition the interval $[a, b]$ into lots of little pieces, or subintervals, on each of which the variable x, and thus the function $f(x)$, does not change that much. The technical phrase for doing this is "to **partition** $[a, b]$," or "to choose a **partition** of $[a, b]$."

DEFINITION 8.1. *A **partition** of $[a, b]$ is a set of points on the x-axis, between $x = a$ and $x = b$, that serve to divide or partition the interval $[a, b]$ into subintervals.*

To specify a partition, just specify the set of points. Typically, the point $x = a$ (the very left-hand endpoint) and the point $x = b$ (the very right-hand endpoint) are included as partition points. Thus, for example, if the original interval is $[2, 4]$, then specifying the partition points

$$2, \quad 2.3, \quad 2.7, \quad 3.1, \quad 3.2, \quad 3.8, \quad 4$$

means that you are dividing or partitioning the interval $[2, 4]$ into subintervals:

$$[2, 2.3], \quad [2.3, 2.7], \quad [2.7, 3.1], \quad [3.1, 3.2], \quad [3.2, 3.8], \quad \text{and} \quad [3.8, 4]$$

Notice that the listing of seven points (including the endpoints 2 and 4) has the result of breaking the interval up into six subintervals. There is a little awkward mismatch between the number of points in the partition and the number of subintervals. For this reason, when counting partition points, the very first one is usually called not the first one, but the zeroth one. Thus in the example above, 2 is the zeroth partition point, 2.3 is the first partition point, and so on. The reason you do this is so that the count of the last partition point is the same as the count of the number of subintervals. In the example above, if 2 is the zeroth partition point, then 4 is the sixth partition point, which matches nicely with the fact that you have six subintervals.

A subscript is usually used to denote the count of a partition point, so that in the example above, 2.7, the second partition point, would be denoted x_2. The fifth partition point, 3.8, would be denoted by x_5. The endpoints of the original interval are also written in the same fashion, so that in the example, $2 = x_0$ and $4 = x_6$.

Putting all the symbols and notation together, you see that a partition of $[a, b]$ is described by a set of points $\{x_0, x_1, x_2, \ldots, x_n\}$, with x_0 standing for the left-hand endpoint a, x_n standing for the right-hand endpoint b, and $a = x_0 < x_1 < x_2 < \ldots < x_n = b$. These points divide $[a, b]$ into subintervals:

The first subinterval is $[x_0, x_1]$.

The second subinterval is $[x_1, x_2]$.

The third subinterval is $[x_2, x_3]$.

If you look at how the pattern goes, you can see that

$$\text{the } i\text{th subinterval is } [x_{i-1}, x_i] \tag{8.2}$$

Also,

$$\text{the } n\text{th and last subinterval is } [x_{n-1}, x_n]$$

The width of the ith interval is found, as for any interval, by computing the difference between the right-hand endpoint of the interval and the left-hand endpoint of the interval, so that

$$\text{the length of the } i\text{th interval } [x_{i-1}, x_i] = x_i - x_{i-1}$$

The abbreviated symbol for this is Δx_i, where Δ is the Greek letter "capital delta." The symbol Δx_i is read as "delta x i" or as "delta x sub i." Thus,

$$\Delta x_i = x_i - x_{i-1} \tag{8.3}$$

What is easiest for computing is to partition the interval $[a, b]$ so that all the subintervals have equal length. Such a partition is called a **regular partition**. Since the length of the whole interval $[a, b]$ is $b - a$, if you divide the whole interval into five equal subintervals, each subinterval will have the length $\dfrac{b-a}{5}$. As all the subintervals in a regular partition have the same length, the symbol Δx is used to denote this length. The subscript i is not necessary for a regular partition. If you divide the whole interval $[a, b]$ into n equal subintervals (n being any positive integer), each subinterval will have the length

$$\Delta x = \frac{b-a}{n} \tag{8.4}$$

With a regular partition, there is a simple formula for the ith partition point x_i. The first partition point is $x_0 = a$. To get to the next partition point x_1, you have to move to the right on the x-axis a distance of $\Delta x = \dfrac{b-a}{n}$. Moving to the right on the number line corresponds to addition. For example, if you are at the point $x = 3$ and you move to the right a distance of 4, you will be at the point $x = 3 + 4 = 7$. Thus moving to the right on the x-axis by a distance of $\Delta x = \dfrac{b-a}{n}$ corresponds to adding $\Delta x = \dfrac{b-a}{n}$. If you start at $x_0 = a$ and move to the right a distance of $\dfrac{b-a}{n}$, you will be adding to $x_0 = a$ the quantity $\dfrac{b-a}{n}$, and you will then be at the point $x_1 = a + 1 \cdot \dfrac{b-a}{n}$. To get to the next partition point x_2, you have to move to the right on the x-axis by *another* distance of $\Delta x = \dfrac{b-a}{n}$, so that

$$x_2 = x_1 + \frac{b-a}{n} = \left(a + \frac{b-a}{n}\right) + \frac{b-a}{n} = a + 2 \cdot \frac{b-a}{n}$$

From the pattern, you can see that the formula for x_i, the right-hand end-point of the ith interval, is

$$x_i = a + i \cdot \frac{b-a}{n} \qquad (8.5)$$

The situation is pictured in Figure 8.7.

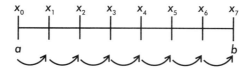

Figure 8.7

You can check that Equation 8.5 gives the right formula for the nth partition point $x_n = b$, as $x_n = a + n \cdot \dfrac{b-a}{n} = a + (b-a) = b$.

Example 8.3 Suppose that the collection of points $\{1, 1.2, 1.4, 2.4, 3, 3.7, 4.2, 4.4, 5\}$ is a partition of the interval $[1,5]$. List all of the subintervals and their lengths.

Solution: Note that the partition of this example is *not* a regular partition.

The first subinterval is $[1, 1.2]$ and has length $\Delta x_1 = 1.2 - 1 = 0.2$.

The second subinterval is $[1.2, 1.4]$ and has length $\Delta x_2 = 1.4 - 1.2 = 0.2$.

The third subinterval is $[1.4, 2.4]$ and has length $\Delta x_3 = 2.4 - 1.4 = 1.0$.

The fourth subinterval is $[2.4, 3]$ and has length $\Delta x_4 = 3 - 2.4 = 0.6$.

The fifth subinterval is $[3, 3.7]$ and has length $\Delta x_5 = 3.7 - 3 = 0.7$.

The sixth subinterval is $[3.7, 4.2]$ and has length $\Delta x_6 = 4.2 - 3.7 = 0.5$.

The seventh subinterval is $[4.2, 4.4]$, and has length $\Delta x_7 = 4.4 - 4.2 = 0.2$.

The eighth and last subinterval is $[4.4, 5]$ and has length $\Delta x_8 = 5 - 4.4 = 0.6$.

Revisiting Problem 1:

Example 8.4 The interval $[0, 2]$ of the bowling ball problem is partitioned by a regular partition into ten subintervals. Find the length of each subinterval, and list the partition points. Also find, in terms of i, the formulas for the endpoints t_{i-1} and t_i of the ith interval.

Solution: Note that as the variable in the bowling ball problem is t for time, you should use t_i and Δt for partition points and lengths of subintervals, rather than use the letter x.

In this example, the interval $[a, b] = [0, 2]$, and the number of subintervals is $n = 10$. By Equation 8.4, the length of each subinterval is

$$\Delta t = \frac{b - a}{n} = \frac{2 - 0}{10} = \frac{2}{10} = 0.2 = \frac{1}{5}$$

Note that $0.2 = \frac{2}{10}$, which reduces to $\frac{1}{5}$. Some people prefer to work with decimals, while others prefer to work with fractions.

You find the partition points by starting at the left-hand endpoint $a = 0$ and adding $\Delta t = 0.2$ each time to get the next partition point, until you end up at $b = 2$. Thus the partition points are

$$\{0, \quad 0.2, \quad 0.4, \quad 0.6, \quad 0.8, \quad 1.0, \quad 1.2, \quad 1.4, \quad 1.6, \quad 1.8, \quad 2.0\}$$

Now t_i, the right-hand endpoint of the ith interval, is found by starting at $a = 0$ and moving to the right by $\Delta t = \dfrac{b - a}{n} = 0.2$ a total of i times, to get, as in Equation 8.5:

$$t_i = a + i \cdot \frac{b - a}{n} = 0 + i \cdot (0.2) = 0.2i = \frac{i}{5}$$

By Equation 8.2, the left-hand endpoint of the ith subinterval is t_{i-1}, which from the above formula (just substitute $i - 1$ for i) is

$$t_{i-1} = 0.2 \cdot (i - 1) = 0.2i - 0.2 = \frac{i - 1}{5}$$

Thus, the ith subinterval is

$$[t_{i-1}, t_i] = [0.2i - 0.2, 0.2i] = \left[\frac{i - 1}{5}, \frac{i}{5}\right]$$

Revisiting Problem 2:

Example 8.5 The interval $[1, 5]$ is partitioned by a regular partition into ten subintervals. Find the length of each subinterval, and list the partition points. Also find, in terms of i, the formulas for the endpoints x_{i-1} and x_i of the ith interval.

Solution: In this example, $[a, b] = [1, 5]$, and the number of subintervals is $n = 10$. By Equation 8.4, the length of each subinterval is

$$\Delta x = \frac{b - a}{n} = \frac{5 - 1}{10} = \frac{4}{10} = \frac{2}{5} = 0.4$$

You find the partition points by starting at the left-hand endpoint $a = 1$ and adding $\Delta x = 0.4$ each time to get the next partition point, until you end up at $b = 5$. Thus the partition points are

$$\{1, \ 1.4, \ 1.8, \ 2.2, \ 2.6, \ 3, \ 3.4, \ 3.8, \ 4.2, \ 4.6, \ 5\}$$

The left-hand endpoint x_i of the ith interval is found by starting at $a = 1$ and moving to the right by $\Delta x = \dfrac{b - a}{n} = 0.4$ a total of i times, to get, as in Equation 8.5:

$$x_i = a + i \cdot \frac{b - a}{n} = 1 + i \cdot (0.4) = 1 + 0.4i = 1 + \frac{2i}{5}$$

By Equation 8.2, the left-hand endpoint of the ith subinterval is x_{i-1}, which from the above formula is

$$x_{i-1} = 1 + (i - 1) \cdot 0.4 = 1 + 0.4(i - 1) = 1 + \frac{2(i - 1)}{5}$$

Thus, the ith subinterval is

$$[x_{i-1}, x_i] = [1 + 0.4(i - 1), \ 1 + 0.4i] = \left[1 + \frac{2(i - 1)}{5}, \ 1 + \frac{2i}{5} \right]$$

Notation and Terminology for "Choosing a Representative or Sample Value in Each Subinterval"

In the bowling ball problem (Problem 1), once you had a small interval of time, you then had to find a representative value for the velocity of the ball in that time interval. This really means finding the value of the velocity function $v(t)$ for *some* value of time t in the small interval. Likewise, for the area problem (Problem 2), you had to find a representative height for each thin vertical strip. The height was simply the y-coordinate for some point on the top of the thin strip, which really means evaluating the function $y = x^2$ for some position x in the small interval.

In both problems, you had to find a representative or sample value for the function on the ith subinterval, and this was done by picking a *representative* or *sample point* p_i in the ith subinterval, and evaluating the function at this point p_i. The point p_i can be *any* particular point in the ith interval, but to make computing easier, you should be systematic about how you choose it. There are generally three ways of choosing p_i systematically: choosing $p_i = x_{i-1}$ to be the left-hand endpoint of the ith subinterval; choosing $p_i = x_i$ to be the right-hand endpoint of the ith subinterval, or choosing $p_i = \dfrac{x_{i-1} + x_i}{2}$ to be the midpoint of the ith subinterval. Below are the formulas for all three choices when $[a, b]$ is partitioned into n equal subintervals. From Equations 8.2 and 8.5, it follows that

1. If p_i is chosen as the left-hand endpoint of the ith subinterval, then

$$p_i = x_{i-1} = a + (i - 1) \cdot \frac{b - a}{n}$$

2. If p_i is chosen as the midpoint of the ith subinterval, then

$$
\begin{aligned}
p_i \ &= \ \frac{x_{i-1} + x_i}{2} \\
&= \ \frac{a + (i - 1) \cdot \frac{b-a}{n} + a + i \cdot \frac{b-a}{n}}{2} \\
&= \ \frac{2a + (2i - 1) \cdot \frac{b-a}{n}}{2} \\
&= \ a + \left(i - \frac{1}{2}\right) \cdot \frac{b - a}{n}
\end{aligned}
$$

3. If p_i is chosen as the right-hand endpoint of the ith subinterval, then

$$p_i = x_i = a + i \cdot \frac{b - a}{n}$$

You can see from the above that choosing p_i to be the right-hand endpoint of each subinterval gives the easiest formulas in terms of i.

Note to students: Many books have problems in which they ask you to choose midpoints or left-hand endpoints in each subinterval. That is why I have included the formulas for these points. However, as the right-hand endpoint has the simplest formula and is easiest to use, most of the examples in this book will use just the right-hand endpoints of each subinterval as sample points.

Example 8.6 For the partition of $[1, 5]$ given in Example 8.3, list the sample points if (a) you choose the left-hand endpoint of each subinterval as a sample point; (b) you choose the right-hand endpoint of each subinterval as a sample point; (c) you choose the midpoint of each subinterval as a sample point.

Solution:

The partition given in Example 8.3 was

$$\{1, \quad 1.2, \quad 1.4, \quad 2.4, \quad 3, \quad 3.7, \quad 4.2, \quad 4.4, \quad 5\},$$

Thus, the subintervals are

$$[1, 1.2], \quad [1.2, 1.4], \quad [1.4, 2.4], \quad [2.4, 3],$$
$$[3, 3.7], \quad [3.7, 4.2], \quad [4.2, 4.4], \quad [4.4, 5]$$

(a) The left-hand endpoints of each subinterval are

$$p_1 = 1, \quad p_2 = 1.2, \quad p_3 = 1.4, \quad p_4 = 2.4,$$
$$p_5 = 3, \quad p_6 = 3.7, \quad p_7 = 4.2, \quad p_8 = 4.4$$

(b) The right-hand endpoints of each subinterval are

$$p_1 = 1.2, \quad p_2 = 1.4, \quad p_3 = 2.4, \quad p_4 = 3,$$
$$p_5 = 3.7, \quad p_6 = 4.2, \quad p_7 = 4.4, \quad p_8 = 5$$

(c) The midpoint p_i of the ith subinterval is the average of the left- and right-hand endpoints of the ith subinterval, so that each p_i is the average of the corresponding p_i's in parts (a) and (b). Thus,

$$p_1 = 1.1, \quad p_2 = 1.3, \quad p_3 = 1.9, \quad p_4 = 2.7,$$
$$p_5 = 3.35, \quad p_6 = 3.95, \quad p_7 = 4.3, \quad p_8 = 4.7$$

Revisiting Problem 2:

Example 8.7 For the regular partition of $[1, 5]$ into ten subintervals given in Example 8.5, list the sample points if you choose (a) the left-hand endpoint of each subinterval as a sample point; (b) the right-hand endpoint of each subinterval as a sample point; (c) the midpoint of each subinterval as a sample point.

Solution: You can work intuitively with the numbers of Example 8.5, or more formally with the formulas for left-hand endpoints, right-hand endpoints, and midpoints given previously. Let's use the formulas to get practice with them. The interval is

$[a, b] = [1, 5]$, and the number of subintervals is $n = 10$, so that by Equation 8.4:

$$\Delta x = \frac{b - a}{n} = \frac{5 - 1}{10} = 0.4 = \frac{2}{5}$$

(a) Using the formula for p_i as the left-hand endpoint,

$$p_i = a + (i - 1)\frac{b - a}{n} = 1 + (i - 1) \cdot 0.4$$

As i goes from 1 to 10, this gives the values

$$
\begin{aligned}
p_1 &= 1 + 0 \cdot (0.4) = 1.0, \\
p_2 &= 1 + 1 \cdot (0.4) = 1.4, \\
p_3 &= 1 + 2 \cdot (0.4) = 1.8
\end{aligned}
$$

Similarly,

$$p_4 = 2.2, \quad p_5 = 2.6, \quad p_6 = 3.0, \quad p_7 = 3.4,$$
$$p_8 = 3.8, \quad p_9 = 4.2, \quad p_{10} = 4.6$$

(b) Using the formula for p_i as the right-hand endpoint,

$$p_i = a + i \cdot \frac{b - a}{n} = 1 + i \cdot (0.4)$$

As i goes from 1 through 10, this gives the values

$$
\begin{aligned}
p_1 &= 1 + 1 \cdot (0.4) = 1.4, \\
p_2 &= 1 + 2 \cdot (0.4) = 1.8, \\
p_3 &= 1 + 3 \cdot (0.4) = 2.2
\end{aligned}
$$

Similarly,

$$p_4 = 2.6, \quad p_5 = 3.0, \quad p_6 = 3.4, \quad p_7 = 3.8,$$
$$p_8 = 4.2, \quad p_9 = 4.6, \quad p_{10} = 5.0$$

(c) Using the formula for p_i as the midpoint,

$$p_i = a + \left(i - \frac{1}{2}\right) \cdot \frac{b - a}{n} = 1 + \left(i - \frac{1}{2}\right) \cdot (0.4)$$

As i goes from 1 through 10, this gives the values

$$
\begin{aligned}
p_1 &= 1 + \left(1 - \frac{1}{2}\right) \cdot (0.4) = 1 + 0.5 \cdot (0.4) = 1 + 0.2 = 1.2 \\
p_2 &= 1 + \left(2 - \frac{1}{2}\right) \cdot (0.4) = 1 + 1.5 \cdot (0.4) = 1 + 0.6 = 1.6 \\
p_3 &= 1 + \left(3 - \frac{1}{2}\right) \cdot (0.4) = 1 + 2.5 \cdot (0.4) = 1 + 1.0 = 2.0
\end{aligned}
$$

Similarly,

$$p_4 = 2.4, \quad p_5 = 2.8, \quad p_6 = 3.2, \quad p_7 = 3.6,$$
$$p_8 = 4.0, \quad p_9 = 4.4, \quad p_{10} = 4.8$$

Notation and Terminology for "Adding Together All the Estimates"

For the bowling ball problem, Problem 1, the estimate for the distance the ball rolls in the ith subinterval was $v(p_i)\Delta t_i$. Here p_i is a particular value of time in the ith time interval, so that $v(p_i)$ is a representative value of the velocity of the bowling ball in that time interval. The representative velocity $v(p_i)$ is multiplied by the length or time duration Δt_i of the time interval to get an estimate of the distance traveled during that time interval. For the area problem, Problem 2, the estimate for the area of the thin vertical strip under the graph of $f(x) = x^2$ and over the ith subinterval was $f(p_i)\Delta x_i = (p_i)^2 \Delta x_i$. This represents an estimate for the height of the thin vertical strip times its width, to give an estimate for the area of the strip. In each case, the estimate for what you wanted to compute, in the ith subinterval, had the *form* $f(p_i)\Delta x_i$.

To get an estimate, or approximate answer, for the whole problem, you just add the estimates for each small interval. Thus the estimated solution for the whole problem has the form

$$f(p_1)\Delta x_1 + \ldots + f(p_n)\Delta x_n \tag{8.6}$$

For the bowling ball problem, this gives an estimate for the distance traveled by the bowling ball. For the area problem, this gives an estimate for the area under the graph. The above type of sum is called a **Riemann sum** or **approximating sum**. Riemann was the German mathematician of the nineteenth century who developed the above ideas in full generality. The term "approximating sum" is also used because the above expression is clearly a sum that approximates what you are trying to compute.

Example 8.8 Estimate the area of the region under the graph of $y = x^2$ and over the x-axis, between $x = 1$ and $x = 5$, by using the partition of $[1, 5]$ given in Example 8.3 and by using the left-hand endpoints of the subintervals as sample or representative points. Also, sketch the region and the approximating rectangles.

Solution:

The widths of the approximating rectangles (see Figure 8.8) are just the widths of the subintervals, as computed in Example 8.3. The heights of the approximating rectangles are found

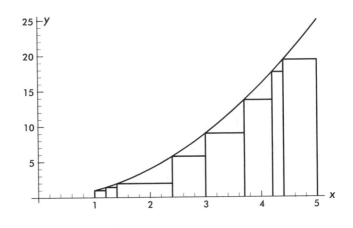

Figure 8.8

simply by evaluating $f(x) = x^2$ at the sample points, which are the left-hand endpoints, as recorded in Example 8.6(a). The areas of the approximating rectangles are computed in the following table:

Interval	Width	Sample point	Height at sample point, using $y = x^2$	Estimated area of interval: area = height · width
$[1, 1.2]$	0.2	1	1	0.2
$[1.2, 1.4]$	0.2	1.2	1.44	0.288
$[1.4, 2.4]$	1	1.4	1.96	1.96
$[2.4, 3]$	0.6	2.4	5.76	3.456
$[3, 3.7]$	0.7	3	9	6.3
$[3.7, 4.2]$	0.5	3.7	13.69	6.845
$[4.2, 4.4]$	0.2	4.2	17.64	3.528
$[4.4, 5]$	0.6	4.4	19.36	11.616

The estimated area is found by adding the estimated areas of all the strips, in the last column, to give 34.193. You can see from the figure that this is an *under*estimate of the true area.

Revisiting Problem 1:

Example 8.9 The bowling ball of Problem 1 has a velocity $v(t) = 9 - 0.2t$ feet per second, t seconds after it is released. Estimate the distance the ball rolls during the first 2 seconds after it is released, by using a regular partition of $[0, 2]$ into ten subintervals, and using for each subinterval the value of the velocity function at the end of the subinterval as a representative value of velocity throughout the subinterval.

Solution: Note first that because in the table on page 159, the velocity was sampled at the midpoint of each interval, the numerical estimate obtained there will be different from the numerical estimate you get in this example. Notice also that unlike in the previous example, the fact that you are using a regular partition makes it easier to begin to organize your work.

You can write the answer in the pattern suggested by Equation 8.6:

$$v(t_1)\Delta t_1 + \ldots + v(t_{10})\Delta t_{10}$$

From Example 8.4, each $\Delta t = 0.2$, and the pattern of right-hand endpoints is given by $0.2, 0.4, 0.6, \ldots, 2.0$ Thus the estimated distance rolled is

$$[v(0.2)] \cdot (0.2) + [v(0.4)] \cdot (0.2) + \ldots + [v(2.0)] \cdot (0.2)$$
$$= (9 - 0.04) \cdot (0.2) + (9 - 0.08) \cdot (0.2) + \ldots + (9 - 0.4) \cdot (0.2)$$
$$= (8.96 + 8.92 + \ldots + 8.60) \cdot (0.2)$$
$$= 87.8 \cdot (0.2) = 17.56$$

Revisiting Problem 2:

Example 8.10 Estimate the area of the region under the graph of $f(x) = x^2$ and over the x-axis, between $x = 1$ and $x = 5$, by using the regular partition of $[1, 5]$ given in Example 8.5, and by using the right-hand endpoints of subintervals as the sample or representative points. Also sketch the region and the approximating rectangles.

Solution: The picture is sketched in Figure 8.9.

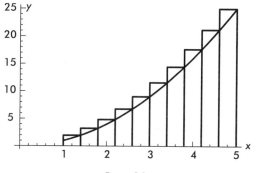

Figure 8.9

The widths of the subintervals and the partition points were computed in Example 8.5.

Note how the heights of the rectangles in each subinterval are now given by the value of $f(x) = x^2$ at the right-hand endpoint of each subinterval, as opposed to Example 8.8, where you were asked to use the left-hand endpoint.

As in the previous example, with a regular partition the work can be organized to conform to the pattern of Equation 8.6. From Example 8.5, the length of each subinterval is $\Delta x = 0.4$, and the sample points are $1.4, 1.8, \ldots, 5$. Thus the estimated area is

$$
\begin{aligned}
& [f(1.4)] \cdot (0.4) + [f(1.8)] \cdot (0.4) + \ldots + [f(5)] \cdot (0.4) \\
= {} & 1.4^2 \cdot (0.4) + 1.8^2 \cdot (0.4) + \ldots + 5^2 \cdot (0.4) \\
= {} & (1.96 + 3.24 + \ldots + 25) \cdot 0.4 \\
= {} & 115.6 \cdot 0.4 = 46.24
\end{aligned}
$$

Summation Notation

The final bit of notation is designed to replace the "$+ \ldots +$" in the Riemann sum with something more explicit and compact, so that you can work with it algebraically. The capital Greek letter sigma, Σ, is called a **summation symbol**, and is used to denote that a whole bunch of terms are to be added together. To use Σ, the different terms to be added together must each depend on an integer, usually denoted by i (j, k, l, m, or n are other common letters used), which increases by 1 from term to term. The integer i is called an **index** or **counter**.

For example, the sum $1 + 4 + 9 + 16 + 25$ equals the sum $1^2 + 2^2 + 3^2 + 4^2 + 5^2$. Thus each term added is of the form i^2, where the index or counter i starts at 1, increases by 1 from term to term, and ends at 5. The letter used for the counter, together with its starting value, is written below the Σ, while the highest value of the counter is written above the Σ. To the right of the Σ is written the formula, as a function of i, for the terms to be added. Thus,

$$
1^2 + 2^2 + 3^2 + 4^2 + 5^2
$$

would be written as

$$
\sum_{i=1}^{5} i^2
$$

The Σ means that a bunch of terms are to be added, and these terms are of the form i^2. The subscript $i = 1$ means that for the first term to be added, you set i equal to 1. You then increase the value of i by 1 each time, and add in another term of the form i^2. The last term to be added in is the term i^2 with the counter i set to the value 5. Thus,

$$\sum_{i=1}^{5} i^2 = 1^2 + 2^2 + 3^2 + 4^2 + 5^2$$
$$= 1 + 4 + 9 + 16 + 25$$
$$= 55$$

Example 8.11 Compute $\displaystyle\sum_{i=1}^{7}(4i - 2)$

Solution:

$$\sum_{i=1}^{7}(4i - 2) = (4 \cdot 1 - 2) + (4 \cdot 2 - 2) + (4 \cdot 3 - 2)$$
$$+ (4 \cdot 4 - 2) + (4 \cdot 5 - 2) + (4 \cdot 6 - 2) + (4 \cdot 7 - 2)$$
$$= 2 + 6 + 10 + 14 + 18 + 22 + 26$$
$$= 98$$

Example 8.12 Compute $\displaystyle\sum_{i=1}^{4}(3i^2 - 2i + 5)$

Solution:

$$\sum_{i=1}^{4}(3i^2 - 2i + 5) = (3 \cdot 1^2 - 2 \cdot 1 + 5) + (3 \cdot 2^2 - 2 \cdot 2 + 5)$$
$$+ (3 \cdot 3^2 - 2 \cdot 3 + 5) + (3 \cdot 4^2 - 2 \cdot 4 + 5)$$
$$= (3 - 2 + 5) + (12 - 4 + 5) + (27 - 6 + 5) + (48 - 8 + 5)$$
$$= 6 + 13 + 26 + 45$$
$$= 90$$

Equation 8.6 for the Riemann or approximating sum was

$$f(p_1)\Delta x_1 + \ldots + f(p_n)\Delta x_n$$

Using the summation symbol Σ, this can be written as

$$\sum_{i=1}^{n} f(p_i)\Delta x_i \tag{8.7}$$

If the interval $[a, b]$ is divided by a regular partition into n equal subintervals, then each subinterval has length $\Delta x = \dfrac{b - a}{n}$. If you choose the representative point p_i in each subinterval to be the right-hand endpoint $x_i = a + i \cdot \dfrac{b-a}{n}$, then the above Riemann sum is

$$\sum_{i=1}^{n} f\left(a + i \cdot \frac{b-a}{n}\right) \cdot \frac{b-a}{n} \tag{8.8}$$

Revisiting Problem 1:

Example 8.13 Write the sum you computed in Example 8.9 for the estimated distance the bowling ball rolls using summation notation.

Solution: As the interval $[0, 2]$ was partitioned into ten equal subintervals, $\Delta t = \dfrac{2-0}{10} = \dfrac{1}{5}$, and the formula for the right-hand endpoint of the ith interval is $p_i = 0 + i \cdot \dfrac{2}{10} = \dfrac{i}{5}$. The velocity function was $v(t) = 9 - 0.2t = 9 - \dfrac{1}{5} \cdot t$. Thus,

$$v(p_i) = v\left(\frac{i}{5}\right) = 9 - \frac{1}{5} \cdot \frac{i}{5} = 9 - \frac{i}{25}$$

So, the sum is

$$\sum_{i=1}^{10} v\left(\frac{i}{5}\right) \cdot \frac{1}{5} = \sum_{i=1}^{10} \left(9 - \frac{i}{25}\right) \cdot \frac{1}{5}$$

Revisiting Problem 2:

Example 8.14 Write, using summation notation, the estimated area computed in Example 8.10.

Solution: Recall that in Example 8.10, $f(x) = x^2$, $a = 1, b = 5$, and $n = 10$, so that $\dfrac{b-a}{n} = 0.4 = \dfrac{2}{5}$ and $p_i = x_i = 1 + i \cdot \dfrac{2}{5} = 1 + \dfrac{2i}{5}$. Thus, for this example, using Equation 8.8 gives

$$\sum_{i=1}^{10} \left(1 + \frac{2i}{5}\right)^2 \cdot \frac{2}{5}$$

How can you get from an estimated answer, involving Riemann sums, to the exact answer to your problem? As before, the idea is that your estimates get better and better as you divide the problem up into smaller and smaller pieces. This is because the smaller each little interval is, the less the function f varies in value over that interval, and the closer f is to being constant. For a regular partition of $[a, b]$, the subintervals become smaller as you choose more of them. This means that you should compute Equations 8.7 or 8.8 over and over again, for larger and larger values of n, getting better and better estimates each time, and see if you can tell what number the estimates are getting closer and closer to. That number will be the exact answer to your problem. Of course, looking at a bunch of numbers and trying to see what value they seem to be getting closer and closer to is precisely the idea of a limit, which you studied in Chapter 3. Thus, the exact answer to your problem will be

$$\lim_{n \to \infty} \sum_{i=1}^{n} f(p_i) \Delta x_i \qquad (8.9)$$

Thus, if you choose the right-hand endpoint x_i as your p_i, your answer will be

$$\lim_{n \to \infty} \sum_{i=1}^{n} f\left(a + i \cdot \frac{b-a}{n}\right) \cdot \frac{b-a}{n} \qquad (8.10)$$

Revisiting Problem 1:

Example 8.15 Write, using limits and summation notation, the exact answer to Problem 1 for the distance rolled by the bowling ball in the first two seconds.

Solution: As $[a, b] = [0, 2]$, $\dfrac{b-a}{n} = \dfrac{2}{n}$, and $a + i \cdot \dfrac{b-a}{n} = i \cdot \dfrac{2}{n} = \dfrac{2i}{n}$.

As $v(t) = 9 - 0.2t$,

$$
\begin{aligned}
v\left(a + i \cdot \frac{b-a}{n}\right) &= v\left(\frac{2i}{n}\right) \\
&= 9 - 0.2 \cdot \left(\frac{2i}{n}\right) \\
&= 9 - \frac{1}{5} \cdot \frac{2i}{n} \\
&= 9 - \frac{2i}{5n}
\end{aligned}
$$

so for Problem 1, Equation 8.10 becomes,

$$\lim_{n \to \infty} \sum_{i=1}^{n} \left(9 - \frac{2i}{5n}\right) \cdot \frac{2}{n} \qquad (8.11)$$

Revisiting Problem 2:

Example 8.16 Write, using limits and summation notation, the exact answer to Problem 2 for the area of the region bounded on the left by $x = 1$, on the right by $x = 5$, below by the x-axis, and above by the curve $f(x) = x^2$.

Solution: As $[a, b] = [1, 5]$, $\dfrac{b-a}{n} = \dfrac{4}{n}$, and $a + i \cdot \dfrac{b-a}{n} = 1 + i \cdot$

$\dfrac{4}{n} = 1 + \dfrac{4i}{n}$. As $f(x) = x^2$, $f(a + i \cdot \dfrac{b-a}{n}) = \left(1 + \dfrac{4i}{n}\right)^2$. Thus for Problem 2, Equation 8.10 becomes

$$\lim_{n \to \infty} \sum_{i=1}^{n} \left(1 + \frac{4i}{n}\right)^2 \cdot \frac{4}{n} \qquad (8.12)$$

You will learn in Chapter 9 how actually to compute limits such as those given in the two previous examples. But first, let's introduce the **integral**.

8.3 THE DEFINITE INTEGRAL: DEFINITION AND NOTATION

This chapter is entitled "The Idea of the Integral," but except for the very beginning of the chapter, the word "integral" hasn't been mentioned. So what is an integral, anyway? Well, if you have a function $f(x)$ defined on an interval $[a, b]$, an integral is the number you get by performing these steps:

1. Partition $[a, b]$ into n little subintervals.
2. Pick a representative point p_i in each subinterval.
3. Compute the Riemann sum $\sum_{i=1}^{n} f(p_i)\Delta x_i$.
4. Keep doing this over and over, letting $n \to \infty$, and compute the limit

$$\lim_{n \to \infty} \sum_{i=1}^{n} f(p_i)\Delta x_i$$

The number representing this limit is called the **definite integral of f from a to b**.

The standard notation for the definite integral of f from a to b is

$$\int_a^b f(x)\, dx$$

In this notation, $\int$ is called the **integral sign**. It is an elongated capital S, of the kind used in the 1600s and 1700s, and which you have probably seen in pictures, for example, of the Declaration of Independence or of the United States Constitution. The letter "S" stands for the "summation" performed in computing a Riemann sum. The numbers a and b are called the **limits of integration**, with the one written on the bottom of the integral sign called the **lower limit of integration** and the one written on the top of the integral sign called the **upper limit of integration**. The limits of integration serve to specify which interval you are partitioning into little subintervals when computing a Riemann sum. The function $f(x)$ to the right of the integral sign is called the **integrand**, and of course represents the function you are evaluating at the representative points p_i. Finally, the term dx is called an **infinitesimal**. The x tells you what the variable of integration is, that is, that you should think of f as a function of x. You can think of the dx as representing what happens to the term Δx in the limit, as the size Δx of the subintervals gets closer and closer to 0. You will recall that the same kind of notation is used for derivatives. Finally, the x is what is called a **dummy variable**. This means that whichever particular letter you use has no importance, and that, for example,

$$\int_a^b x^2 \, dx = \int_a^b t^2 \, dt$$

In each case, you are taking the "squaring function" on the interval from a to b, and for that function on that interval, you are computing a limit of Riemann sums.

To summarize,

DEFINITION 8.2. *The definite integral of f from a to b,*

$$\int_a^b f(x) \, dx$$

is defined as $\lim_{n \to \infty} \sum_{i=1}^n f(p_i) \Delta x_i$, *where the interval $[a, b]$ is partitioned into n subintervals, Δx_i is the length of the ith subinterval, p_i is some representative or sample value of x in the ith subinterval, and as the number n of subintervals increases to ∞, you choose the partition points so that the length of the largest subinterval in each partition is approaching zero.*

You will recall from Chapter 3, "The Idea of Limits," that a limit does not always exist. The following theorem, which we will not prove, guarantees that in one very common case, the limit does exist.

THEOREM 8.1. *Let the function $f(x)$ be continuous on the interval $[a, b]$. Then the limit used in the definition of $\int_a^b f(x) \, dx$ exists.*

Idea of the Proof: Continuous functions are precisely those functions with the property that, if you look at them on subintervals that are getting thinner and thinner, then the values of the function vary by less and less. The basic idea of the integral was that if you looked at a function on a small subinterval, then the values did not vary by very much on that subinterval, and you could make the estimate that the function was nearly constant on the small subinterval. Thus it is precisely for the collection of continuous functions that the basic idea of the integral is valid.

Revisiting Problem 1:

Example 8.17 Write, using the notation of the integral, the expression for the distance the bowling ball of Problem 1 on page 155 rolls during the first two seconds.

> **Solution:** As the time interval is $[0, 2]$ and the velocity function is $v(t) = 9 - 0.2t$, the answer is
>
> $$\int_0^2 (9 - 0.2t)\, dt$$

Revisiting Problem 2:

Example 8.18 Write, using integral notation, the expression for the area of the region bounded on the left by $x = 1$, on the right by $x = 5$, below by the x-axis, and above by the curve $y = x^2$.

> **Solution:** As the interval is $[1, 5]$, and the function is $f(x) = x^2$, the answer is
>
> $$\int_1^5 x^2\, dx$$

8.4 SUMMARY OF MAIN POINTS

- Problems that can be easily solved when all of the problem ingredients are constant can often be solved when the problem ingredients are variable.

- More specifically, if you are given a function $f(x) = C$ that is constant on an interval $[a, b]$, and the solution to your problem is computed by multiplying the constant value C of the function by the length $b - a$ of the interval, then the same problem can often be solved when the function $f(x)$ is variable.

  — One common example involves computing the distance an object moves when you know its velocity function $v(t)$. If t stands for time, and the velocity of the object is constant, $v(t) = C$, then from time $t = a$ to time $t = b$, that is, over a time interval of duration $b - a$, the distance the object travels equals $C \cdot (b - a)$. For constant velocity, distance equals velocity times time elapsed.

  — The second common example involves the area of the region in the plane between the x-axis, the graph of a positive function $f(x)$, and two vertical lines, one at $x = a$ and the other at $x = b$. If the function $f(x)$ is constant, so that $f(x) = C$, then the region is just a rectangle of height C and width $b - a$, so that the area equals $C \cdot (b - a)$. The area of a rectangle equals its height times its width.

- When the function $f(x)$ is variable, the method is guaranteed to work provided the function $f(x)$ is *continuous* on the interval $[a, b]$.

- The idea of the method is the following:

  — Partition or subdivide the interval $[a, b]$ into lots of small subintervals. For many problems, and in particular for Problems 1 and 2, the solution on the whole interval is simply the *sum* of the solutions of the corresponding problem on the subintervals.

  — On a *small* subinterval of $[a, b]$, the function $f(x)$ does not vary very much, and is therefore approximately constant. Pick one of the values of $f(x)$ on the subinterval to serve as a representative value on the subinterval, and multiply the representative value by the width of the subinterval. This gives an approximate or estimated solution for the problem on the subinterval.

  — Add the estimated solutions for all the subintervals to get an estimated or approximate solution for the full problem. This estimate is called a Riemann sum, or an approximate sum.

— Repeat the above steps over and over, choosing more and more
subintervals that are getting smaller and smaller. The numbers you
get are better and better estimates or approximations to the exact
answer.

— The exact answer is the *limit* of the above estimates.

• Summation notation is useful for writing Riemann or approximate sums.
Using summation notation, a sum such as $a_1 + a_2 + \ldots + a_n$ can be written
as $\sum_{i=1}^{n} a_i$.

• The standard notation for keeping track of all of the above is the following:

— The division of the interval $[a, b]$ into subintervals can be described
by giving the set of *partition points* $\{x_0, x_1, \ldots, x_n\}$. The partition points are the endpoints of the subintervals, and the leftmost
endpoint a is also written as x_0, while the rightmost endpoint
b is also written as x_n. The first subinterval is $[a, x_1] = [x_0, x_1]$.
The second subinterval is $[x_1, x_2]$. Continuing with this pattern, the
ith subinterval is $[x_{i-1}, x_i]$, while the nth and last subinterval is
$[x_{n-1}, x_n] = [x_{n-1}, b]$. Again, with this notation, there are n subintervals.

— The notation p_i is used to stand for the point in the ith interval at
which to evaluate f. Thus $f(p_i)$ is the representative value of f on
the ith subinterval.

— The width of the ith subinterval is $x_i - x_{i-1}$. The shorthand for this
is Δx_i.

— With all of this notation, the approximate sum is

$$\sum_{i=1}^{n} f(p_i) \Delta x_i$$

The exact answer is

$$\lim_{n \to \infty} \sum_{i=1}^{n} f(p_i) \Delta x_i$$

• For actually computing an answer by the above method, the following
choices are the most convenient:

— Partition $[a, b]$ with a regular partition, where all the subintervals
have the same size. With n subintervals, the size of each one is
simply $\Delta x = \dfrac{b-a}{n}$. Furthermore, x_i, the right-hand endpoint of
the ith interval, is given by

$$x_i = a + i \cdot \frac{b-a}{n} = a + i \cdot \Delta x$$

You can see this because you get to the right-hand endpoint of the ith subinterval by starting at a and moving to the right by $\Delta x = \dfrac{b-a}{n}$ a total of i times.

— It is most convenient to pick as p_i the point $x_i = a + i \cdot \dfrac{b-a}{n}$.

— With these choices, the Riemann sum is

$$\sum_{i=1}^{n} f\left(a + i \cdot \frac{b-a}{n}\right) \cdot \frac{b-a}{n}$$

The exact answer is given by

$$\lim_{n\to\infty} \sum_{i=1}^{n} f\left(a + i \cdot \frac{b-a}{n}\right) \cdot \frac{b-a}{n}$$

- The number you get from doing all of the above is called "the definite integral of f from a to b."

- The definite integral of f from a to b is written symbolically as

$$\int_a^b f(x)\,dx$$

— The $\int$ is called the integral sign.

— The quantity a is called the lower limit of integration, while b is called the upper limit of integration.

— The function $f(x)$ is called the integrand.

— The variable x is a dummy variable, in the sense that $\int_a^b f(x)\,dx = \int_a^b f(t)\,dt$. It does not matter what letter you use.

8.5 EXERCISES

In Exercises 1–6, you are asked to estimate the area under the graph of $y = x^2 + x$ and over the x-axis, between $x = 0$ on the left and $x = 3$ on the right, by using the given partition of $[0, 3]$ and the given sample points in each subinterval. For each problem, sketch the curve *and* the approximating rectangles.

1. Use the partition $P = \{0,\ 0.4,\ 0.8,\ 1.6,\ 2.4,\ 2.9,\ 3\}$, with the left-hand endpoint of each subinterval as a sample point.

2. Use the same partition as in Exercise 1, with the midpoint of each subinterval as a sample point.

3. Use the same partition as in Exercise 1, with the right-hand endpoint of each subinterval as a sample point.

4. Use a regular partition into six subintervals, with the left-hand endpoint of each subinterval as a sample point.

5. Use a regular partition into six subintervals, with the midpoint of each interval as a sample point.

6. Use a regular partition into six subintervals, with the right-hand endpoint of each subinterval as a sample point.

Exercises 7–11 still deal with the area under the graph of $y = x^2 + x$ and over the interval $[0, 3]$ on the x-axis, but deal more with notation than with the actual numerical computation of an estimated area.

7. In Exercise 6, write, in terms of i, the formula for the right-hand endpoint of the ith subinterval. Then write out the answer to Exercise 6 using summation notation.

8. Using a regular partition of $[0, 3]$ into 100 subintervals, and the right-hand endpoint of each subinterval as a sample point, write out, using summation notation, the formula for the estimated area under the graph of $y = x^2 + x$.

9. Do the same as in Exercise 8, but with a regular partition of $[0, 3]$ into n subintervals.

10. Write the formula for the *exact* area under the graph of $y = x^2 + x$ and over the interval $[0, 3]$, using limits and summation notation.

11. Write the formula for the same area as in Exercise 10, using the notation of the integral.

In Exercises 12–13, evalute the following summations.

12. $\sum_{i=1}^{6} (3i - 5)$

13. $\sum_{i=1}^{4} (6i^2 - 3i + 4)$

Rewrite the expressions in Exercises 14–16 using summation notation. You do not have to evaluate them.

14. $3 \cdot 1^2 + 3 \cdot 2^2 + \dots + 3 \cdot 97^2$

15. $2 + 4 + 6 + \dots + 20$

16. $1 + 3 + 5 + 7 + \dots + 37$

Each of the integrals in Exercises 17–19 can be interpreted as the area of a region in the plane. In each case, first sketch the region, and then write the formula for the area of the region as a limit of Riemann or approximate sums (that is, using limit and summation notation).

17. $\displaystyle\int_0^2 (x^2 + 1)\, dx$

18. $\displaystyle\int_1^3 2t\, dt$

19. $\displaystyle\int_{-2}^3 (x^2 + 4)\, dx$

20. A ball thrown down from the roof of a building at a speed of 10 feet per second has velocity equal to $v(t) = -32t - 10$ feet per second, t seconds after it is thrown. Assuming the ball has not yet hit the ground 3 seconds after it was thrown, estimate the total distance traveled by the ball during the first 3 seconds after it was thrown, by subdividing the time interval $[0, 3]$ into six equal subintervals and sampling the velocity at the end of each time subinterval.

9
COMPUTING SOME INTEGRALS

In this chapter you will learn, for some relatively simple functions, how actually to compute the limit of a Riemann sum, as given by Equation 8.10 of Section 8.2. The following chapter contains an easier alternate approach. There are three reasons for working through the more difficult computations of this chapter. The first is simply to see that it can be done, and the second is to be able to better appreciate, by comparison, the easier approach of the next chapter. Neither of these reasons is very compelling. The most important reason for working through the computations of this chapter is to become more familiar and at ease with the *definition* of the integral as a limit of Riemann or approximate sums. It is *only* through the definition that the use of the integral to solve problems has meaning. Why, for example, should the integral of x^2 from $x = 1$ to $x = 5$ represent the area of the region under the graph of x^2 and over the x-axis, between the vertical lines $x = 1$ and $x = 5$? It is only because of the procedure of the last chapter, leading up to the definition of the integral, that the answer is clear. The easier procedure of the next chapter for *computing* the integral of x^2 from $x = 1$ to $x = 5$ gives absolutely no clue whatsoever as to why the value of the integral represents, for example, the area of a region in the plane. It is only by *combining* the definition of the integral, which provides the intuition as to why the integral can be used to solve problems, with the easier method of computing integrals given in the next chapter, that the integral has become such a powerful and pervasive tool throughout mathematics and all of the sciences.

9.1 SUMMATION RULES AND FORMULAS

The rules and formulas given in this section for dealing with sums provide the tools needed to change the computing of a Riemann sum from an exercise in arithmetic to more of an exercise in algebra. This allows you to do two things: You can compute fairly easily Riemann sums where the number of subintervals is large. You can also get compact, manageable formulas for the sum when the number of subintervals is left unspecified, as n, so that you can investigate what happens as $n \to \infty$. The summation rules are nothing more than the usual rules of arithmetic and algebra, rewritten in the Σ notation. One simple rule is

$$\sum_{i=1}^{n} c = nc \tag{9.1}$$

Here, the terms being added together are constants c that do not depend upon the counter i. As the counter i goes from 1 to n, a total of n c's are being added together, to give nc. A specific example would be

$$\sum_{i=1}^{7} 3 = 3+3+3+3+3+3+3 = 7 \cdot 3 = 21$$

In the next three equations, terms such as a_i or b_i stand for any expressions that depend upon i. Thus a_i could be i^2, or b_i could be $2i - 3$. If $a_i = i^2$, then $a_1 = 1^2 = 1, a_2 = 2^2 = 4$, and so on.

$$\sum_{i=1}^{n} ca_i = c \sum_{i=1}^{n} a_i \tag{9.2}$$

This is just the distributive law of arithmetic, and says that

$$ca_1 + \ldots + ca_n = c(a_1 + \ldots a_n)$$

A specific example would be

$$\sum_{i=1}^{4} 3i^2 = 3 \sum_{i=1}^{4} i^2$$

This holds because, clearly,

$$3 \cdot 1^2 + 3 \cdot 2^2 + 3 \cdot 3^2 + 3 \cdot 4^2 = 3 \cdot (1^2 + 2^2 + 3^2 + 4^2)$$

For the summation of a sum of two indexed terms,

$$\sum_{i=1}^{n} (a_i + b_i) = \sum_{i=1}^{n} a_i + \sum_{i=1}^{n} b_i \tag{9.3}$$

This is just the commutative law of addition, and says that

$$(a_1 + b_1) + \ldots + (a_n + b_n) = (a_1 + \ldots + a_n) + (b_1 + \ldots + b_n)$$

A specific example would be

$$\sum_{i=1}^{4} (i + i^2) = \sum_{i=1}^{4} i + \sum_{i=1}^{4} i^2$$

This holds because, clearly,

$$(1 + 1^2) + (2 + 2^2) + (3 + 3^2) + (4 + 4^2) = (1 + 2 + 3 + 4) + (1^2 + 2^2 + 3^2 + 4^2)$$

Similarly,

$$\sum_{i=1}^{n}(a_i - b_i) = \sum_{i=1}^{n} a_i - \sum_{i=1}^{n} b_i \tag{9.4}$$

The next set of formulas are not merely symbolic expressions of the elementary rules of arithmetic, but rather are results with significant content. There is no need to complicate matters here by giving proofs of these rules—you can probably find their proofs in your regular calculus text. In any case, you can certainly check out for particular examples that the rules work, if you want to convince yourself emotionally that the rules hold. These rules will allow you to express summations in a more concise form, so that you can more easily compute them.

Note to students: Ask your professor if he or she wants you to memorize the three following equations.

First,

$$\sum_{i=1}^{n} i = \frac{n(n + 1)}{2} \tag{9.5}$$

For $n = 5$, this formula says that $\sum_{i=1}^{5} i = 1 + 2 + 3 + 4 + 5$ can also be computed as $\dfrac{5 \cdot (5 + 1)}{2} = \dfrac{5 \cdot 6}{2} = 15$, which indeed equals the above sum.

Second,

$$\sum_{i=1}^{n} i^2 = \frac{n(n + 1)(2n + 1)}{6} \tag{9.6}$$

For $n = 5$, this formula says that

$$\sum_{i=1}^{5} i^2 = 1^2 + 2^2 + 3^2 + 4^2 + 5^2 = 1 + 4 + 9 + 16 + 25 = 55$$

can also be computed as

$$\frac{5 \cdot (5 + 1) \cdot (2 \cdot 5 + 1)}{6} = \frac{5 \cdot 6 \cdot 11}{6} = 55$$

Finally,

$$\sum_{i=1}^{n} i^3 = \frac{n^2(n + 1)^2}{4} \tag{9.7}$$

For $n = 5$, this says that

$$\sum_{i=1}^{5} i^3 = 1^3 + 2^3 + 3^3 + 4^3 + 5^3 = 1 + 8 + 27 + 64 + 125 = 225$$

can also be computed as

$$\frac{5^2 \cdot (5+1)^2}{4} = \frac{25 \cdot 36}{4} = 25 \cdot 9 = 225$$

Example 9.1 Use the summation rules and formulas to reevaluate the summation $\sum_{i=1}^{7}(4i-2)$ of Example 8.11 of Section 8.2.

Solution:

$$
\begin{aligned}
\sum_{i=1}^{7}(4i-2) &= \sum_{i=1}^{7}4i - \sum_{i=1}^{7}2 && \text{(Eq. 9.4)}\\
&= 4\sum_{i=1}^{7}i - 7 \cdot 2 && \text{(Eqs. 9.2 and 9.1)}\\
&= 4 \cdot \left(\frac{7 \cdot 8}{2}\right) - 14 && \text{(Eq. 9.5)}\\
&= 112 - 14 = 98
\end{aligned}
$$

Example 9.2 Use the summation rules and formulas to reevaluate the summation $\sum_{i=1}^{4}(3i^2 - 2i + 5)$ of Example 8.12 of Section 8.2.

Solution:

$$
\begin{aligned}
\sum_{i=1}^{4}(3i^2 - 2i + 5) &= \sum_{i=1}^{4}3i^2 - \sum_{i=1}^{4}2i + \sum_{i=1}^{4}5 && \text{(by Eqs. 9.3 and 9.4)}\\
&= 3\sum_{i=1}^{4}i^2 - 2\sum_{i=1}^{4}i + 4 \cdot 5 && \text{(Eqs. 9.2 and 9.1)}\\
&= 3 \cdot \left(\frac{4 \cdot 5 \cdot 9}{6}\right) - 2 \cdot \left(\frac{4 \cdot 5}{2}\right) + 20 && \text{(Eqs. 9.6 and 9.5)}\\
&= 3 \cdot 30 - 2 \cdot 10 + 20 = 90
\end{aligned}
$$

Revisiting Problem 1 of Section 8.1:

Example 9.3 Use the summation rules and formulas to compute the summation of Example 8.9 of Section 8.2 to estimate how far the bowling ball rolled in 2 seconds.

Solution: Example 8.13 of Section 8.2 shows that the sum of Example 8.9 can be written in summation notation as

$$\sum_{i=1}^{10} \left(9 - \frac{i}{25}\right) \cdot \frac{1}{5}$$

To use the summation rules and formulas you must multiply out the expression, in order to separate out the parts that do not depend upon i from the parts that do. As $(9 - \frac{i}{25}) \cdot \frac{1}{5} = \frac{9}{5} - \frac{i}{125}$,

$$
\begin{aligned}
\sum_{i=1}^{10} \left(9 - \frac{i}{25}\right) \cdot \frac{1}{5} &= \sum_{i=1}^{10} \left(\frac{9}{5} - \frac{i}{125}\right) \\
&= \sum_{i=1}^{10} \frac{9}{5} - \sum_{i=1}^{10} \frac{i}{125} && \text{(Eq. 9.4)} \\
&= \frac{90}{5} - \frac{1}{125} \sum_{i=1}^{10} i && \text{(Eq. 9.1 and 9.2)} \\
&= 18 - \frac{1}{125} \cdot \left(\frac{10 \cdot 11}{2}\right) && \text{(Eq. 9.5)} \\
&= 18 - \frac{55}{125} = 18 - 0.44 = 17.56
\end{aligned}
$$

This is of course the same answer you found before, in Chapter 8. However, with hardly any more work, you could use the methods of this chapter to get an even better estimate by partitioning the interval into 100 subintervals. Doing this purely numerically would be quite a chore.

Example 9.4 Use the summation rules and formulas to estimate the distance the bowling ball of Problem 1 of Section 8.1 rolls in the first two seconds. Do this by partitioning the interval into 100 subintervals of equal length, and estimate the velocity during each subinterval by using the velocity of the ball at the end of the subinterval.

Solution: Use Equation 8.8 of Section 8.2 for the Riemann sum:

$$\sum_{i=1}^{n} f\left(a + i \cdot \frac{b-a}{n}\right) \cdot \frac{b-a}{n}$$

In this example, the interval $[a, b] = [0, 2]$, and the number of subintervals is $n = 100$, so that $\frac{b-a}{n} = \frac{2-0}{100} = \frac{1}{50}$. Also,

$$a + i \cdot \frac{b-a}{n} = \frac{i}{50}, \text{ and } v(t) = 9 - \frac{1}{5} \cdot t, \text{ so that}$$

$$v\left(\frac{i}{50}\right) = 9 - \frac{1}{5} \cdot \frac{i}{50} = 9 - \frac{i}{250}$$

Thus the terms to be summed are of the form

$$\left(9 - \frac{i}{250}\right) \cdot \frac{1}{50}$$

As in the previous example, you must multiply out the expression to separate out the parts that do not depend upon i from those that do before you can use the summation rules and formulas. As

$$\left(9 - \frac{i}{250}\right) \cdot \frac{1}{50} = \frac{9}{50} - \frac{i}{12,500}$$

you must evaluate

$$\sum_{i=1}^{100} \left(\frac{9}{50} - \frac{i}{12,500}\right) = \sum_{i=1}^{100} \frac{9}{50} - \sum_{i=1}^{100} \frac{i}{12,500} \qquad \text{(Eq. 9.4)}$$

$$= \frac{900}{50} - \frac{1}{12,500} \sum_{i=1}^{100} i \qquad \text{(Eqs. 9.1 and 9.2)}$$

$$= 18 - \frac{1}{12,500} \cdot \left(\frac{100 \cdot 101}{2}\right) \qquad \text{(Eq. 9.5)}$$

$$= 18 - \frac{5,050}{12,500}$$

$$= 18 - 0.404 = 17.596$$

The next example divides the interval $[0, 2]$ into n subintervals.

Example 9.5 Use the summation rules and formulas to estimate the distance rolled by the bowling ball during the first two seconds. Get your estimate by partitioning the interval $[0, 2]$ into n equal subintervals, and choose the right-hand endpoint of each subinterval as your representative point.

Solution: Use Equation 8.8 of Section 8.2 again:

$$\sum_{i=1}^{n} f\left(a + i \cdot \frac{b-a}{n}\right) \cdot \frac{b-a}{n}$$

In this example, $[a, b] = [0, 2]$ and $v(t) = 9 - \dfrac{1}{5} \cdot t$ as before. The number of subintervals is simply n, so that $\dfrac{b - a}{n} = \dfrac{2 - 0}{n} = \dfrac{2}{n}$, and $a + i \cdot \dfrac{b - a}{n} = i \cdot \dfrac{2}{n} = \dfrac{2i}{n}$. Thus,

$$v\left(a + i \cdot \frac{b - a}{n}\right) = v\left(\frac{2i}{n}\right) = 9 - \frac{1}{5} \cdot \frac{2i}{n} = 9 - \frac{2i}{5n}$$

So, the terms to be summed are of the form

$$\left(9 - \frac{2i}{5n}\right) \cdot \frac{2}{n} = \frac{18}{n} - \frac{4i}{5n^2}$$

The Riemann sum is

$$
\begin{aligned}
\sum_{i=1}^{n} \left(\frac{18}{n} - \frac{4i}{5n^2}\right) &= \frac{18}{n} \sum_{i=1}^{n} 1 - \frac{4}{5n^2} \sum_{i=1}^{n} i && \text{(Eqs. 9.4 and 9.2)} \\
&= \frac{18}{n} \cdot n - \frac{4}{5n^2} \cdot \frac{n(n+1)}{2} && \text{(Eqs. 9.1 and 9.5)} \\
&= 18 - \frac{2}{5} \cdot \frac{n(n+1)}{n^2}
\end{aligned}
$$

You can now check Examples 9.3 and 9.4 simply by substituting $n = 10$, and then $n = 100$, into the formula just found. You can also see that with virtually no more work, you could get more and more accurate estimates by evaluating this formula for larger and larger n.

Revisiting Problem 2 of Section 8.1:

Example 9.6 Use the summation rules and formulas to compute the summation of Example 8.10 of Section 8.2.

Solution: Example 8.14 of Section 8.2 shows that the sum of Example 8.10 can be written in summation notation as

$$\sum_{i=1}^{10} \left(1 + \frac{2i}{5}\right)^2 \cdot \frac{2}{5}$$

The summation rules and formulas can't be used until you separate out the parts of the above equation that do not depend at all on i from those that have the term i in them, and from those that have the term i^2 in them. To do this, you must first expand the square and then multiply by $\frac{2}{5}$ to get

$$\left(1 + \frac{2i}{5}\right)^2 \cdot \frac{2}{5} = \left(1 + \frac{4i}{5} + \frac{4i^2}{25}\right) \cdot \frac{2}{5} = \frac{2}{5} + \frac{8i}{25} + \frac{8i^2}{125}$$

The Riemann sum is

$$\sum_{i=1}^{10} \left(\frac{2}{5} + \frac{8i}{25} + \frac{8i^2}{125}\right)$$

$$= \sum_{i=1}^{10} \frac{2}{5} + \sum_{i=1}^{10} \frac{8i}{25} + \sum_{i=1}^{10} \frac{8i^2}{125} \qquad \text{(Eq. 9.3)}$$

$$= \frac{2}{5}\sum_{i=1}^{10} 1 + \frac{8}{25}\sum_{i=1}^{10} i + \frac{8}{125}\sum_{i=1}^{10} i^2 \qquad \text{(Eq. 9.2)}$$

$$= \frac{2}{5}\cdot 10 + \frac{8}{25}\cdot\left(\frac{10\cdot 11}{2}\right) + \frac{8}{125}\cdot\left(\frac{10\cdot 11\cdot 21}{6}\right) \qquad \text{(Eqs. 9.1, 9.5, and 9.6)}$$

$$= \frac{2}{5}\cdot 10 + \frac{8}{25}\cdot 55 + \frac{8}{125}\cdot 385$$

$$= \frac{100 + 440 + 616}{25} = \frac{1,156}{25} = 46\frac{6}{25} = 46.24$$

This is of course the same answer as obtained from the numerical work of Example 8.10.

The next example will involve getting a better estimate by partitioning the interval into many more subintervals. You'll recall that in the discussion of Problem 2 of Section 8.1, increasing the number of subintervals was mentioned as a possibility, but the numerical work seemed overwhelming. By converting the work from numerical work to algebraic work, it is hardly any more difficult. The example after that will involve dividing the original interval into n subintervals. With a general, rather than a specific, number of subintervals, you will be able to see what happens in the limit as the number n of subintervals gets larger and larger, that is, as $n \to \infty$.

Example 9.7 Use the summation rules and formulas to estimate the area of the region in the xy-plane bounded below by the x-axis, on the left by the line $x = 1$, on the right by the line $x = 5$, and above by the curve $f(x) = x^2$. Get your estimate by partitioning the interval $[1, 5]$ into 100 equal subintervals, and choose the right-hand endpoint of each subinterval as your representative point.

Solution: Use Equation 8.8 of Section 8.2:

$$\sum_{i=1}^{n} f\left(a + i \cdot \frac{b-a}{n}\right) \cdot \frac{b-a}{n}$$

In this example, $[a, b] = [1, 5]$ and $n = 100$, so that

$$\frac{b-a}{n} = \frac{5-1}{100} = \frac{4}{100} = \frac{1}{25}$$

Also,

$$a + i \cdot \frac{b-a}{n} = 1 + i \cdot \frac{1}{25} = 1 + \frac{i}{25}$$

As $f(x) = x^2$,

$$f\left(a + i \cdot \frac{b-a}{n}\right) = f\left(1 + \frac{i}{25}\right) = \left(1 + \frac{i}{25}\right)^2$$

So, the terms to be summed are of the form

$$\left(1 + \frac{i}{25}\right)^2 \cdot \frac{1}{25} = \left(1 + \frac{2i}{25} + \frac{i^2}{625}\right) \cdot \frac{1}{25}$$

$$= \frac{1}{25} + \frac{2i}{625} + \frac{i^2}{15,625}$$

The Riemann sum is

$$\sum_{i=1}^{100} \left(\frac{1}{25} + \frac{2i}{625} + \frac{i^2}{15,625}\right)$$

$$= \sum_{i=1}^{100} \frac{1}{25} + \sum_{i=1}^{100} \frac{2i}{625} + \sum_{i=1}^{100} \frac{i^2}{15,625} \qquad \text{(Eq. 9.3)}$$

$$= \frac{1}{25} \sum_{i=1}^{100} 1 + \frac{2}{625} \sum_{i=1}^{100} i + \frac{1}{15,625} \sum_{i=1}^{100} i^2 \qquad \text{(Eq. 9.2)}$$

$$= \frac{1}{25} \cdot 100 + \frac{2}{625} \cdot \left(\frac{100 \cdot 101}{2}\right)$$

$$+ \frac{1}{15,625} \cdot \left(\frac{100 \cdot 101 \cdot 201}{6}\right) \qquad \text{(Eqs. 9.1, 9.5, and 9.6)}$$

$$= \frac{1}{25} \cdot 100 + \frac{2}{625} \cdot (5,050) + \frac{1}{15,625} \cdot (338,350)$$

$$= \frac{2,500 + 10,100 + 13,534}{625} = \frac{26,134}{625} = 41.8144$$

Notice that when you use the summation rules and formulas, computing the approximate sum after partitioning the interval into 100 equal subintervals is not that much harder than computing it after partitioning the interval into ten equal subintervals—at least if you have a calculator. If you just worked *completely* numerically, as we did in constructing the table at the beginning of Chapter 8, then even with a calculator, increasing the number of subintervals from 10 to 100 would have been much, much more work. Surprisingly enough, when you use the summation rules and formulas and you replace a large n, such as $n = 100$, with a *general* n, for the purpose of computing the limit as $n \to \infty$, the work becomes even easier.

Example 9.8 Use the summation rules and formulas to estimate the area of the region in the xy-plane bounded below by the x-axis, on the left by the line $x = 1$, on the right by the line $x = 5$, and above by the curve $f(x) = x^2$. Get your estimate by partitioning the interval $[1, 5]$ into n equal subintervals, and choose the right-hand endpoint of each subinterval as your representative point.

Solution: Once more use Equation 8.8 of Section 8.2:

$$\sum_{i=1}^{n} f\left(a + i \cdot \frac{b-a}{n}\right) \cdot \frac{b-a}{n}$$

In this example, $[a, b] = [1, 5]$ and $f(x) = x^2$, as before, but the number of subintervals is simply n. Thus,

$$\frac{b-a}{n} = \frac{5-1}{n} = \frac{4}{n}$$

Also,

$$a + i \cdot \frac{b-a}{n} = 1 + i \cdot \frac{4}{n} = 1 + \frac{4i}{n}$$

It follows that

$$f\left(a + i \cdot \frac{b-a}{n}\right) = f\left(1 + \frac{4i}{n}\right) = \left(1 + \frac{4i}{n}\right)^2$$

So, the terms to be summed are of the form

$$\left(1 + \frac{4i}{n}\right)^2 \cdot \frac{4}{n} = \left(1 + \frac{8i}{n} + \frac{16i^2}{n^2}\right) \cdot \frac{4}{n}$$

$$= \frac{4}{n} + \frac{32i}{n^2} + \frac{64i^2}{n^3}$$

The Riemann sum is therefore

$$\sum_{i=1}^{n} \left(\frac{4}{n} + \frac{32i}{n^2} + \frac{64i^2}{n^3} \right)$$

$$= \frac{4}{n} \sum_{i=1}^{n} 1 + \frac{32}{n^2} \sum_{i=1}^{n} i + \frac{64}{n^3} \sum_{i=1}^{n} i^2 \qquad \text{(Eqs. 9.3 and 9.2)}$$

$$= \frac{4}{n} \cdot n + \frac{32}{n^2} \cdot \frac{n(n+1)}{2} + \frac{64}{n^3} \cdot \frac{n(n+1)(2n+1)}{6} \qquad \text{(Eqs. 9.1, 9.5, and 9.6)}$$

$$= 4 + 16 \cdot \frac{n(n+1)}{n^2} + \frac{64}{6} \cdot \frac{n(n+1)(2n+1)}{n^3}$$

Note that you can evaluate the formula just found at $n = 10$ or at $n = 100$ to check the estimates in Examples 9.7 and 9.8. You could also evaluate the formula at $n = 1,000$ or at $n = 1,000,000$ to get even better estimates. But more is true—you can compute the limit as $n \to \infty$ of the above formula to get the *exact* answer for the area under the graph.

9.2 COMPUTING LIMITS OF APPROXIMATE SUMS

The Riemann sums computed in Examples 9.5 and 9.8 involve expressions of the form $\frac{p(n)}{q(n)}$, where $p(n)$ and $q(n)$ are polynomials in n. The following small subsection will tell you how to find limits of expressions of this form, and after that, you will be able to compute limits of Riemann or approximate sums.

Some More Limits

The type of limit you must be able to compute is of the form $\lim_{n \to \infty} \frac{p(n)}{q(n)}$, where $p(n)$ and $q(n)$ are polynomials in n. There is a general principle that makes computing this limit easy. The idea is that, for large values of n, the value of the highest power of n in a polynomial will have much more influence on the value of the polynomial than the values of all the other terms. For example, if you look at $p(n) = n^3 - 5n^2 + 4n - 7$, and evaluate it at a large value of n, say $n = 9,000$, the value of the n^3-term overwhelms the values of the other terms combined, even though the other terms have larger coefficients. Get out your calculator and check it! Suppose you had another cubic polynomial like $q(n) = 2n^3 + 7n^2 - 4n + 6$. As the value of $p(n)$ at $n = 9,000$ is mostly determined by the highest order term n^3 in $p(n)$, and as the value of $q(n)$ at $n = 9,000$ is also mostly determined by the highest order term $2n^3$ in $q(n)$, it stands to reason that the value of the ratio

$$\frac{p(n)}{q(n)} = \frac{n^3 - 5n^2 + 4n - 7}{2n^3 + 7n^2 - 4n + 6}$$

for large values of n is mostly determined by the ratios of the high order terms:

$$\frac{n^3}{2n^3} = \frac{1}{2}$$

Thus, for large values of n, $\frac{p(n)}{q(n)}$ should be about $\frac{1}{2}$, and accordingly,

$$
\begin{aligned}
\lim_{n\to\infty} \frac{p(n)}{q(n)} &= \lim_{n\to\infty} \frac{n^3 - 5n^2 + 4n - 7}{2n^3 + 7n^2 - 4n + 6} \\
&= \lim_{n\to\infty} \frac{n^3}{2n^3} = \frac{1}{2}
\end{aligned}
$$

Again, get out your calculator and compute $\frac{p(n)}{q(n)}$. You don't even have to try n as big as 9,000. Try computing it for $n = 1{,}000$.

The above is an intuitive explanation for the following theorem. There is a more formal algebraic proof also.

THEOREM 9.1. *Let $p(n)$ and $q(n)$ be polynomials. Then*

$$\lim_{n\to\infty} \frac{p(n)}{q(n)} = \lim_{n\to\infty} \frac{\text{highest order term of } p(n)}{\text{highest order term of } q(n)}$$

Revisiting Problem 1 of Section 8.1:

Example 9.9 Use the summation rules and formulas, together with limits, to compute the exact distance rolled by the bowling ball of Problem 1 in Section 8.1 during the first two seconds.

Solution: By Example 9.5, the approximate sum estimate for the distance rolled when the interval is partitioned into n equal subintervals is

$$18 - \frac{2}{5} \cdot \frac{n(n+1)}{n^2}$$

Thus the exact answer is given by

$$
\begin{aligned}
\lim_{n\to\infty} \left(18 - \frac{2}{5} \cdot \frac{n(n+1)}{n^2} \right) &= \lim_{n\to\infty} 18 - \frac{2}{5} \lim_{n\to\infty} \frac{n(n+1)}{n^2} \\
&= 18 - \frac{2}{5} \lim_{n\to\infty} \frac{n^2}{n^2} \\
&= 18 - \frac{2}{5} \cdot 1 = 17.6
\end{aligned}
$$

Revisiting Problem 2 of Section 8.1:

Example 9.10 Use the summation rules and formulas, together with limits, to compute the exact area of the region in the xy-plane bounded below by the x-axis, on the left by the line $x = 1$, on the right by the line $x = 5$, and above by the curve $f(x) = x^2$.

Solution: By Example 9.8, the approximate sum estimate of the area when the interval $[1, 5]$ is partitioned into n equal subintervals is

$$4 + 16 \cdot \frac{n(n+1)}{n^2} + \frac{64}{6} \cdot \frac{n(n+1)(2n+1)}{n^3}$$

Thus the exact answer is given by

$$\lim_{n \to \infty} \left(4 + 16 \cdot \frac{n(n+1)}{n^2} + \frac{64}{6} \cdot \frac{n(n+1)(2n+1)}{n^3} \right)$$

$$= \lim_{n \to \infty} 4 + 16 \lim_{n \to \infty} \frac{n(n+1)}{n^2} + \frac{64}{6} \lim_{n \to \infty} \frac{n(n+1)(2n+1)}{n^3}$$

$$= 4 + 16 \lim_{n \to \infty} \frac{n^2}{n^2} + \frac{64}{6} \lim_{n \to \infty} \frac{2n^3}{n^3}$$

$$= 4 + 16 \cdot 1 + \frac{64}{6} \cdot 2 = 41 \frac{1}{3}$$

Example 9.11 Evaluate $\displaystyle\int_0^3 (x^2 - 2x + 3)\, dx$

Solution: The answer will represent the area of the region in the plane under the graph of $f(x) = x^2 - 2x + 3$ and over the x-axis between the vertical lines $x = 0$ and $x = 3$. Using Definition 8.2 of Section 8.3 for the integral, the easiest way to proceed is by partitioning $[0, 3]$ into n *equal* subintervals and picking the *right-hand endpoint* of each subinterval as a point at which to evaluate $f(x) = x^2 - 2x + 3$. Then compute the Riemann sum, simplify it using the summation rules and formulas, and finally evaluate the limit as $n \to \infty$.

As $[a, b] = [0, 3]$,

$$\frac{b - a}{n} = \Delta x = \frac{3}{n}$$

The right-hand endpoint of each subinterval is

$$x_i = \left(a + i \cdot \frac{b - a}{n} \right) = \frac{3i}{n}$$

Thus,

$$
\begin{aligned}
\int_0^3 (x^2 - 2x + 3)\, dx &= \lim_{n\to\infty} \sum_{i=1}^n f\left(\frac{3i}{n}\right)\frac{3}{n} \\
&= \lim_{n\to\infty} \sum_{i=1}^n \left[\left(\frac{3i}{n}\right)^2 - 2\cdot\frac{3i}{n} + 3\right]\frac{3}{n} \\
&= \lim_{n\to\infty} \sum_{i=1}^n \left(\frac{9i^2}{n^2} - \frac{6i}{n} + 3\right)\frac{3}{n} \\
&= \lim_{n\to\infty} \sum_{i=1}^n \left(\frac{27i^2}{n^3} - \frac{18i}{n^2} + \frac{9}{n}\right) \\
&= \lim_{n\to\infty} \left(\sum_{i=1}^n \frac{27i^2}{n^3} - \sum_{i=1}^n \frac{18i}{n^2} + \sum_{i=1}^n \frac{9}{n}\right) \\
&= \lim_{n\to\infty} \left(\frac{27}{n^3}\sum_{i=1}^n i^2 - \frac{18}{n^2}\sum_{i=1}^n i + \frac{9}{n}\sum_{i=1}^n 1\right) \\
&= \lim_{n\to\infty} \left(\frac{27}{n^3}\cdot\frac{n(n+1)(2n+1)}{6} - \frac{18}{n^2}\cdot\frac{n(n+1)}{2} + \frac{9}{n}n\right) \\
&= \lim_{n\to\infty} \left(\frac{27\cdot 2n^3}{6n^3} - \frac{18\cdot n^2}{2n^2} + \frac{9n}{n}\right) \\
&= \frac{27\cdot 2}{6} - \frac{18}{2} + 9 = 9 - 9 + 9 = 9
\end{aligned}
$$

9.3 SUMMARY OF MAIN POINTS

- For actually computing a limit of Riemann sums, the following choices are the most convenient.

  — Partition $[a, b]$ with a regular partition, where all the subintervals have the same size. With n subintervals, the size of each one is simply $\Delta x = \dfrac{b-a}{n}$. Furthermore, x_i, the right-hand endpoint of the ith subinterval, is given by

  $$
  x_i = a + i\cdot\frac{b-a}{n} = a + i\cdot\Delta x
  $$

  You can see this because you get to the right-hand endpoint of the ith subinterval by starting at a and moving to the right by $\Delta x = \dfrac{b-a}{n}$ a total of i times.

— It is most convenient to pick as p_i the point $x_i = a + i \cdot \dfrac{b-a}{n}$.

— With these choices, the Riemann sum is

$$\sum_{i=1}^{n} f\left(a + i \cdot \frac{b-a}{n}\right) \cdot \frac{b-a}{n}$$

The exact answer is given by

$$\lim_{n \to \infty} \sum_{i=1}^{n} f\left(a + i \cdot \frac{b-a}{n}\right) \cdot \frac{b-a}{n}$$

- To actually compute a number from the formula above, you can use the summation rules and formulas to rewrite the Riemann sum so that the letter i disappears. It is then possible to compute the limit as $n \to \infty$.

  — The first step is to expand $f\left(a + i \cdot \dfrac{b-a}{n}\right)$ so that you can separate out the terms that do not depend upon i from the terms with i in them, and from the terms with i^2 or something else in them.

  — The next step is to use the summation rules to rework your expression so that the only parts that depend explicitly upon i are of the form $\sum_{i=1}^{n} 1$ or $\sum_{i=1}^{n} i$ or $\sum_{i=1}^{n} i^2$, and so on.

  — Then use the summation formulas to replace the above terms with, respectively, n or $\dfrac{n(n+1)}{2}$ or $\dfrac{n(n+1)(2n+1)}{6}$, and so on.

  — By now your expression should have no Σ's in it. Finally, use the rule that if $p(n)$ and $q(n)$ are polynomials in n, then

  $$\lim_{n \to \infty} \frac{p(n)}{q(n)} = \lim_{n \to \infty} \frac{\text{the highest order term in } p(n)}{\text{the highest order term in } q(n)}$$

  to evaluate the limit of the Riemann or approximate sums.

- The number you get from doing all of the above is called "the definite integral of f from a to b."

- The definite integral of f from a to b is written symbolically as

$$\int_{a}^{b} f(x)\, dx$$

9.4 EXERCISES

Use the summation rules and formulas to evaluate the sums in Exercises 1–4.

1. $\displaystyle\sum_{i=1}^{8}(3i+7)$

2. $\displaystyle\sum_{i=1}^{5}(2i^2+4i-6)$

3. $\displaystyle\sum_{i=1}^{999}(7i-3)$

4. $\displaystyle\sum_{i=1}^{38}(i^2-i+2)$

Use the summation rules and formulas to evaluate the sums in Exercises 5 and 6, in terms of n, as simply as you can.

5. $\displaystyle\sum_{i=1}^{n}(4-3i)$

6. $\displaystyle\sum_{i=1}^{n}(i^2+3i-2)$

Evaluate the limits in Exercises 7–9.

7. $\displaystyle\lim_{n\to\infty}\frac{n^3-2n^2+3n-4}{7n^2+5n-6}$

8. $\displaystyle\lim_{n\to\infty}\frac{3n^2-4n+6}{2n^2+9n-8}$

9. $\displaystyle\lim_{n\to\infty}\frac{7n^2+5n-3}{n^3+3n^2-2n-1}$

Use the summation rules and formulas to evaluate the limits in Exercises 10 and 11.

10. $\displaystyle\lim_{n\to\infty}\sum_{i=1}^{n}\left[2(\frac{3i}{n})^2+\frac{3i}{n}\right]\cdot\frac{3}{n}$

11. $\displaystyle\lim_{n\to\infty}\sum_{i=1}^{n}\left[7(1+\frac{2i}{n})-4\right]\cdot\frac{2}{n}$

12. Find the exact area under the graph of $y = x^2 + x + 1$ and over the x-axis, between the y-axis on the left and the vertical line $x = 2$ on the right. Do this by subdividing the interval $[0, 2]$ into n equal subintervals, computing a Riemann or approximate sum, and then computing the limit of the approximate sums as $n \to \infty$.

13. A ball thrown up from a building with an initial velocity of 64 feet per second has velocity $v(t) = -32t + 64$ feet per second, t seconds after it is thrown. Find the exact distance traveled by the ball during the first two seconds after it is thrown by representing the distance as a limit of Riemann or approximate sums.

14. Evaluate $\int_0^4 (x^2 - 3x)\, dx$ by using the definition of the integral as a limit of Riemann or approximate sums.

10

FORMULAS FOR INTEGRALS: INTEGRALS, ANTIDERIVATIVES, AND THE FUNDAMENTAL THEOREM OF CALCULUS

10.1 INTRODUCTION

In the previous two chapters, you learned about the idea of the integral, and what is meant by computing the definite integral of a function $f(x)$ over the interval $[a, b]$. It looks like actually doing the computation is a pretty tough job. But recall your earlier work with derivatives. After the idea of the derivative and the definition of the derivative of a function $f(x)$, were introduced, it turned out that there were rules and procedures, easier to apply than the definition, for computing derivatives. The same holds for integrals. There are procedures for computing the definite integral of a function $f(x)$ over the interval $[a, b]$ that are easier than applying the definition. As with derivatives, however, the definition is important because it is only through the definition that you can understand why the integral gives the answer to particular problems.

The easier procedure for computing definite integrals relates integrals to derivatives. This relationship is so important in calculus that the theorem that describes the relationship is called the Fundamental Theorem of Calculus. In the next two sections, you will see a statement of what the Fundamental Theorem of Calculus (FTC) says, and an idea of why the theorem is true. Sections 10.4 and 10.5 will show you how to use the Fundamental Theorem of Calculus to compute some definite integrals.

10.2 THE FUNDAMENTAL THEOREM OF CALCULUS—THE MAIN IDEA

In Chapter 8, one problem that motivatived the introduction of the idea of the integral was the problem of finding the area of the region in the xy-plane bounded below by the x-axis, on the left by the line $x = 1$, on the right by the line $x = 5$, and above by the curve $y = x^2$. Let's consider a somewhat different problem. Suppose now, instead of picking a particular right-hand boundary for the region whose area you want to find (such as $x = 5$ above), you want to be a little fancier. Let's say you wanted to get a formula you could just substitute into, so that if I gave you the right-hand boundary as $x = 6$, you could just substitute 6 into the formula to get the area, and likewise if I gave you $x = 4$ as the right-hand boundary, you could just substitute 4 into the formula to get the area. In other words, you want a function, in terms of the variable x, that will represent the area under the graph of $y = x^2$ and over the x-axis, between $x = 1$ on the left and a general vertical line at position x on the right.

There's a little problem in the last sentence. The variable x is being used for two different things. It is being used as the name of a "generic" point on the horizontal axis, and as the name of the variable position of the right-hand vertical boundary line. This can be fixed up by changing the name of the letter we use to represent one of these two things. By tradition, and also because it will make the moral of the whole story stand out more clearly, the name x will be kept for the position of the right-hand vertical boundary line, and the "generic" point on the horizontal axis will be called t. The horizontal axis will now be the t-axis.

The problem is to find the area of the region in the ty-plane bounded below by the t-axis, on the left by the line $t = 1$, on the right by the (variable) vertical line $t = x$, and above by the graph of the function $y = t^2$ (see Figure 10.1).

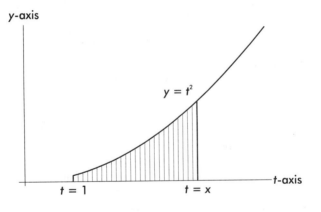

Figure 10.1

The answer should be a function of x. Using the ideas of the previous chapter on computing areas by limits of Riemann, or approximate, sums, and the definite integral notation for a limit of Riemann sums, the problem can be expressed symbolically as finding a formula, in terms of x, for

$$\int_1^x t^2 \, dt$$

In order to stress the main point without getting bogged down in too much algebra, it will be helpful to actually look at an easier problem, where the upper boundary of the area is given by the function or equation $y = t$. Thus you will be computing, as a function of x, the area of the region shown in Figure 10.2.

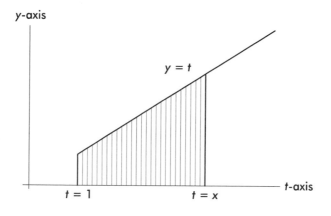

y-axis

$y = t$

$t = 1$

$t = x$

t-axis

Figure 10.2

Symbolically, you will be computing, as a function of x,

$$\int_1^x t \, dt$$

Recall from the previous two chapters what the above integral means. You take the interval from 1 to x on the t-axis, which has length equal to $x - 1$, and divide it into n equal subintervals, each of which has length $\Delta t = \dfrac{x - 1}{n}$. (I am using Δt instead of Δx for the name of a small subinterval because I changed the name of the horizontal axis from the x-axis to the t-axis.) As in Chapter 8, the ith subinterval can be denoted by $[t_{i-1}, t_i]$, where the formula for t_i, the right-hand endpoint of the ith subinterval, is obtained by adding to the left-hand endpoint $t = 1$ the length of i subintervals, so that

$$t_i = 1 + i\Delta t = 1 + i\left(\frac{x - 1}{n}\right)$$

You must pick a sample point p_i in the ith subinterval at which to evaluate the function $f(t)$, and as before it is easiest to pick p_i as the right-hand endpoint t_i, so that

$$p_i = 1 + i\left(\frac{x - 1}{n}\right)$$

The limit of the Riemann sums is then

$$\lim_{n \to \infty} \sum_{i=1}^n f(p_i)\Delta t \quad = \quad \lim_{n \to \infty} \sum_{i=1}^n p_i \Delta t$$

$$= \lim_{n \to \infty} \left[\sum_{i=1}^{n} \left[1 + i \left(\frac{x-1}{n} \right) \right] \left(\frac{x-1}{n} \right) \right]$$

$$= \lim_{n \to \infty} \sum_{i=1}^{n} \left[\left(\frac{x-1}{n} \right) + i \left(\frac{x-1}{n} \right)^2 \right]$$

$$= \lim_{n \to \infty} \left[\sum_{i=1}^{n} \left(\frac{x-1}{n} \right) + \sum_{i=1}^{n} i \left(\frac{x-1}{n} \right)^2 \right]$$

$$= \lim_{n \to \infty} \left[\frac{x-1}{n} \sum_{i=1}^{n} 1 + \left(\frac{x-1}{n} \right)^2 \sum_{i=1}^{n} i \right]$$

$$= \lim_{n \to \infty} \left[\frac{x-1}{n} n + \left(\frac{x-1}{n} \right)^2 \frac{n(n+1)}{2} \right]$$

$$= \lim_{n \to \infty} \left[(x-1) + (x-1)^2 \left(\frac{n^2+n}{2n^2} \right) \right]$$

$$= (x-1) + (x-1)^2 \cdot \lim_{n \to \infty} \frac{n^2+n}{2n^2}$$

$$= (x-1) + (x-1)^2 \cdot \lim_{n \to \infty} \frac{n^2}{2n^2}$$

$$= (x-1) + \frac{(x-1)^2}{2}$$

$$= (x-1) + \frac{x^2 - 2x + 1}{2}$$

$$= \frac{1}{2}x^2 - \frac{1}{2}$$

Remember that the above function, $\frac{1}{2}x^2 - \frac{1}{2}$, gives the area under the graph of $f(t) = t$ and over the t-axis, between the vertical line $t = 1$ on the left and the vertical line $t = x$ on the right. As the function represents an area, a good name for it is $A(x)$. Thus,

$$A(x) = \frac{1}{2}x^2 - \frac{1}{2} = \int_1^x t \, dt$$

There are two important things to notice about this function $A(x)$. The first is that $A(1) = \frac{1}{2} \cdot 1 - \frac{1}{2} = 0$. This makes sense because $A(1)$ is supposed to be the area under the graph of $y = t$ and over the t-axis, between $t = 1$ and $t = 1$. In other words, the region whose area you are computing collapses in the case $x = 1$ to a vertical line segment, and the area of a line segment is 0. In fact, the integral $\int_a^a f(x) \, dx$ is *always* defined to be 0, no matter what the function

$f(x)$ is, for the same reason. The interval over which you are integrating the function $f(x)$ is the interval $[a, a]$. That is, the interval has really collapsed to a point, and the integral has the value 0. Thus,

$$\int_a^a f(x)\, dx = 0 \tag{10.1}$$

The second important thing to notice about $A(x)$ is that if you take its derivative, you get $A'(x) = \frac{1}{2} \cdot 2x = x$. Aside from the fact that you switched the name of the variable from x to t, you can see that at least in this example, the derivative of the area function gives back the original integrand, that is, the original function you integrated. Thus,

$$\frac{d}{dx}\left(\int_1^x t\, dt\right) = x \tag{10.2}$$

Suppose it were true for *any* function $f(t)$ and *any* value of $t = a$ that if the "area" function is defined as

$$A(x) = \int_a^x f(t)\, dt$$

then the derivative of the "area" function would give back the original integrand f. Suppose it were true that

$$A'(x) = \frac{d}{dx}\left(\int_a^x f(t)\, dt\right) = f(x) \tag{10.3}$$

Could that help you evaluate $\int_a^b f(t)\, dt$? The answer is that it could, and you will now see how. Suppose that $B(x)$ is *any* function whose derivative also equals $f(x)$, so that

$$B'(x) = A'(x) = f(x)$$

Then $A(x)$ and $B(x)$ have the same derivative, and thus the function $B - A$ has derivative equal to 0, because

$$(B - A)'(x) = B'(x) - A'(x) = f(x) - f(x) = 0$$

It follows from Theorem 6.5 of Section 6.2 that the function $B - A$ is therefore a constant, so that

$$(B - A)(x) = B(x) - A(x) = C$$

Thus there is a constant or number C such that

$$B(x) = A(x) + C, \text{ for all } x$$

Now by the definition of $A(x) = \int_a^x f(t)\, dt$, and the fact, mentioned above, that $A(a) = 0$, it follows that

$$\int_a^b f(t)\, dt = A(b) = A(b) - A(a)$$

Let's compute $B(b) - B(a)$:

$$
\begin{aligned}
B(b) - B(a) &= [A(b) + C] - [A(a) + C] \\
&= A(b) + C - A(a) - C \\
&= A(b) - A(a)
\end{aligned}
$$

This is precisely the value of $\int_a^b f(t)\, dt$.

Thus for *any* function $B(x)$ whose derivative $B'(x) = f(x)$, it follows that you can evaluate $\int_a^b f(t)\, dt$ by simply computing $B(b) - B(a)$. If you remember the comments about dummy variables at the end of Chapter 8, you will remember that there is absolutely no difference between $\int_a^b f(t)\, dt$ and $\int_a^b f(x)\, dx$. Thus, if all of the above were true, then for any function $f(x)$, you could compute $\int_a^b f(x)\, dx$ simply by finding a function $B(x)$ whose derivative $B'(x) = f(x)$, and simply computing $B(b) - B(a)$.

There is one theoretical point that is required for all of the above indeed to hold. What is required is the hypothesis that the function $f(x)$ be *continuous* on the interval $[a, b]$ over which you are integrating. The next section will give you an idea of why the derivative of the "area" function equals the integrand, and why the assumption that the integrand is continuous is necessary.

SOME TERMINOLOGY: *A function* $B(x)$ *whose derivative equals* $f(x)$, *so that*

$$
B'(x) = f(x)
$$

is called an **anti-derivative** *of* $f(x)$.

The Fundamental Theorem of Calculus relates integrals to derivatives, and says that the easy way to compute a definite integral is not by computing a limit of Riemann sums, but by finding an antiderivative.

The part of the Fundamental Theorem of Calculus that will be most useful for you is the following:

THEOREM 10.1. *If the function* $f(x)$ *is continuous on the interval* $[a, b]$, *then to compute* $\int_a^b f(x)\, dx$ *simply find an antiderivative* $F(x)$ *of* $f(x)$, *that is, a function* $F(x)$ *such that* $F'(x) = f(x)$, *and compute* $F(b) - F(a)$.

10.3 THE FUNDAMENTAL THEOREM OF CALCULUS—AN IDEA OF THE PROOF

In this section, you'll see the idea behind the proof of the FTC.

> ### THE FUNDAMENTAL THEOREM OF CALCULUS
>
> **Part I:** Let $f(t)$ be a continuous function on the interval $[a, b]$. Then the function $A(x)$, defined by the formula
>
> $$A(x) = \int_a^x f(t)\, dt \text{ for all } x \text{ in } [a, b]$$
>
> is an antiderivative of $f(x)$, meaning that
>
> $$A'(x) = \frac{d}{dx}\left(\int_a^x f(t)\, dt\right) = f(x) \text{ for all } x \text{ in } [a, b]$$
>
> **Part II:** Let $F(x)$ be *any* antiderivative of $f(x)$ on $[a, b]$, so that
>
> $$F'(x) = f(x) \text{ for all } x \text{ in } [a, b]$$
>
> Then
>
> $$\int_a^b f(x)dx = F(b) - F(a)$$

You have already seen, in the previous section, an explanation of why the second part of the Fundamental Theorem of Calculus follows from the first. This second part is the part you will use most often, in your computation of definite integrals.

What do you have to do to prove Part I? Recall the definition of the derivative, Definition 2.2 of Section 2.3. By the definition of the derivative, you must show that

$$\lim_{h \to 0} \frac{A(x + h) - A(x)}{h} \tag{10.4}$$

exists, and that this limit equals $f(x)$.

So that you can get a picture of what's going on, let's assume that the function $f(x)$ has positive values on the interval $[a, b]$, so that an integral can be thought of as representing an area under the graph of f. You should think of x as representing some point between a and b, and of h as representing some small positive number, so that $x + h$ is also between a and b, and is a little bit to the right of x. On the number line, the picture is as shown in Figure 10.3.

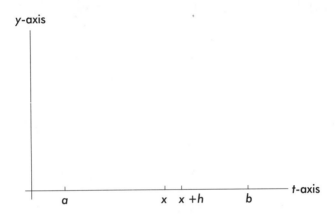

Figure 10.3

Then from the definition of $A(x)$,

$$A(x + h) = \int_a^{x+h} f(t)\, dt$$

This represents the area of the region under the graph of f and over the horizontal axis, bounded on the left by the vertical line $t = a$ and on the right by the vertical line $t = x + h$, as pictured in Figure 10.4.

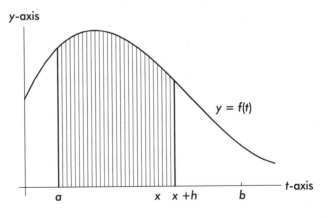

Figure 10.4. $A(x + h)$

Likewise,

$$A(x) = \int_a^x f(t)\, dt$$

represents the area of a similar region, but bounded on the right by the vertical line $t = x$, as shown in Figure 10.5.

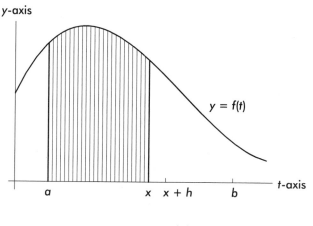

Figure 10.5. $A(x)$

Now the numerator of Equation 10.4 is simply the difference $A(x+h) - A(x)$. If you subtract one area from the other, you clearly get the area under the graph of f and above the horizontal axis, bounded on the left by the vertical line $t = x$ and on the right by the vertical line $t = x + h$, as pictured in Figure 10.6.

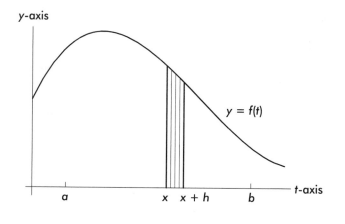

Figure 10.6. $A(x + h) - A(x)$

Now look at this area. Does it remind you of anything? Hopefully it reminds you of the kind of *thin strip* whose area we tried to approximate in Chapter 8.

Now the top of the strip is given by the graph of a *continuous* function, so that the values of the function do not vary very much in the little interval. It follows that the thin strip is approximately a rectangle, and its area can be estimated by multiplying the width by an estimate for the height. As in Chapter 8, the width is simply

$$(x + h) - x = h$$

The height can be estimated by picking *any* sample point p in the interval $[x, x + h]$ and using $f(p)$ for a "representative" height. Thus, for p any value in the interval $[x, x + h]$,

$$A(x + h) - A(x) \text{ approximately equals } f(p) \cdot h$$

Looking again at Equation 10.4, you see that

$$\frac{A(x + h) - A(x)}{h} \text{ approximately equals } \frac{f(p) \cdot h}{h} = f(p).$$

Finally, as $h \to 0$, the point p, which is sandwiched in between x and $x + h$, gets closer and closer to x, and again by *continuity* of f, the values $f(p)$ get closer and closer to the value $f(x)$. Thus, under the assumption that the function f is continuous on the interval $[a, b]$, it is indeed true that

$$\lim_{h \to 0} \frac{A(x + h) - A(x)}{h} = f(x)$$

This completes the "informal" proof of the Fundamental Theorem of Calculus.

Note to students: Some calculus books attempt to present a somewhat more rigorous proof of the Fundamental Theorem of Calculus, by arguing that the area of the strip is *exactly* equal to $f(p) \cdot h$ for *some* point p in the interval $[x, x + h]$. This argument is based upon the Intermediate Value Theorem, as well as several other theorems that cannot be rigorously proven within the confines of a standard first-year calculus book.

10.4 COMPUTING SOME ANTIDERIVATIVES

Now that you know that integrals are computed by finding antiderivatives, the question is "how do you find antiderivatives?" The best answer is that you should know the rules and formulas for finding derivatives *very* well, so that you can recognize patterns, and be able to use the derivative rules and formulas *backwards*.

As the key first step in finding a definite integral $\int_a^b f(x)\,dx$ is finding an antiderivative of $f(x)$, this first step has its own terminology and notation.

DEFINITION 10.1. *The **indefinite integral** of $f(x)$, denoted by*

$$\int f(x)\,dx$$

without limits of integration, is simply the general antiderivative of $f(x)$.

Recall from Section 10.2 that if $F(x)$ is *one* antiderivative of $f(x)$; then the general antiderivative has the form $F(x) + C$, where C is what is called "an arbitrary constant," that is, any constant whatsoever. In finding the indefinite integral, or general antiderivative, of a function, you should find one particular antiderivative, and then add the arbitary constant. Thus, if for a particular f you are asked to evaluate $\int f(x)\, dx$, your answer should always have the form

$$\int f(x)\, dx = F(x) + C$$

THEOREM 10.2.

1. (a) $\displaystyle\int 0 \, dx = C$

    (b) *The integral or antiderivative of 0 is a constant.*

2. (a) $\displaystyle\int 1 \, dx = x + C$

    (b) *The integral of the constant 1 is the variable of integration.*

Proof: Of course, the theorem is true because the derivative of a constant equals 0, and because the derivative of x with respect to x is just 1.

For another example of working a derivative rule backwards, recall the power rule for derivatives:

$$\frac{d}{dx}(x^n) = nx^{n-1}$$

In words, you take the derivative of x to a power by multiplying in front by the exponent and subtracting 1 from the exponent. Thus $\frac{d}{dx}(x^3) = 3x^2$. Suppose you were asked to find $\int 3x^2 \, dx$. That means you want to find an antiderivative of $3x^2$, a function $F(x)$ whose derivative is $3x^2$. Well, you see one a few lines above—it is just x^3. Thus,

$$\int 3x^2 \, dx = x^3 + C$$

How could you have found the integral if you didn't have the answer essentially staring you in the face? Well, analyze the power rule for derivatives: you multiply by the exponent and subtract one from the exponent. The first step in working the rule backwards would say that you *add* one to the exponent. That's one clue as to how to get x^3 as an antiderivative of $3x^2$: add 1 to the exponent of 2 to get an exponent of 3, and the term x^3.

Another way to see essentially the same thing is to look for patterns: when you take the derivative of a power, you get another power, but a power that is smaller by 1. A term like x^2 comes up when you take the derivative of x^3, a term like x^6 comes up when you take the derivative of x^7, and so on. So how could you get a term like x^6 as the derivative of something else? The answer is by taking the derivative of one power higher, namely, x^7. That is, because derivatives *lower* the power by one, antiderivatives should *raise* the power by one.

If you try taking the derivative of x^7, however, you don't get exactly x^6, you get $7x^6$. Your answer for the antiderivative is 7 times as big as it should be. You can rectify this by dividing your answer by 7. You can get a term like x^6 as the derivative of something else by letting the something else be $\frac{1}{7}x^7$. Now it should be clear what the second step in making the power rule work backwards should be: divide by the new exponent.

THEOREM 10.3. *The Power Rule for Integrals.*

1.

$$\int x^n \, dx = \frac{1}{n+1} x^{n+1} + C \text{ for } n \neq -1$$

2. *To compute the antiderivative of a variable raised to a power, add 1 to the power, and divide by the new power.*

The case where the power is -1 to begin with is an exception: if $n = -1$, then $\frac{1}{n+1}$ is undefined.

There is a slight variant on the above approach to the power rule for integrals. If you can correctly guess the *pattern* of an antiderivative, and check your answer by differentiating, and find out that you are all right as far as the part of the function with the variable is concerned, but are off by a multiplicative constant, you can always fix things up to get the constant right. For example, if you want to find an antiderivative of x^4, and see that you should add 1 to the power and come up with a tentative guess of x^5, then you can always check your guess. As $\frac{d(x^5)}{dx} = 5x^4$, your guess is too big by a factor of 5. You can remedy this by multiplying your initial guess by $\frac{1}{5}$, and change your guess from x^5 to $\frac{1}{5}x^5$. Then, because the derivative of a constant times a function equals the constant times the derivative of the function, you get

$$\left(\frac{1}{5}x^5\right)' = \frac{1}{5} \cdot (x^5)' = \frac{1}{5} \cdot 5x^4 = x^4$$

This is exactly what you want the derivative to be. Thus, $\int x^4 \, dx = \frac{1}{5}x^5 + C$.

As with derivatives, it is often convenient to convert expressions involving square root signs or fractional expressions to exponent form so that you can use the power rule for integrals.

Example 10.1 Find $\int x^{12} \, dx$.

Solution: Just use the power rule for integrals to get

$$\int x^{12}\, dx = \frac{1}{13}x^{13} + C$$

Example 10.2 Find $\int t^{12}\, dt$.

Solution: Remember that the particular letter used for the variable does not matter. The previous example asked for a function of x whose derivative with respect to x was x^{12}. This example asks for a function of t whose derivative with respect to t is t^{12}. The function t^{12} is still a power of the variable, and the power rule for integrals applies. Thus,

$$\int t^{12}\, dt = \frac{1}{13}t^{13} + C$$

Example 10.3 Find $\int \frac{1}{x^3}\, dx$.

Solution: Rewrite the integrand $\frac{1}{x^3}$ as x^{-3} and use the power rule for integrals. Remember that the exponent is now -3, and that if you raise it by 1, you get $-3+1 = -2$. You still divide by the new exponent, as before. Thus,

$$\begin{aligned}
\int \frac{1}{x^3}\, dx &= \int x^{-3}\, dx \\
&= \frac{1}{-2}x^{-2} + C \\
&= -\frac{1}{2}x^{-2} + C = -\frac{1}{2x^2} + C
\end{aligned}$$

Example 10.4 Find $\int \sqrt{t}\, dt$.

Solution: Exactly as with derivatives, rewrite $\sqrt{t}$ as $t^{\frac{1}{2}}$, and use the power rule for integrals, to get

$$\int \sqrt{t}\, dt = \int t^{\frac{1}{2}}\, dt = \frac{1}{\frac{3}{2}}t^{\frac{3}{2}} + C = \frac{2}{3}t^{\frac{3}{2}} + C$$

Remember, dividing by $\frac{3}{2}$ is the same as multiplying by its reciprocal $\frac{2}{3}$.

Example 10.5 Find $\displaystyle\int \frac{1}{\sqrt{x}}\, dx$.

Solution:

$$\int \frac{1}{\sqrt{x}}\, dx = \int \frac{1}{x^{\frac{1}{2}}}\, dx$$
$$= \int x^{-\frac{1}{2}}\, dx$$
$$= 2x^{\frac{1}{2}} + C$$

Do you see where the exponent of $\frac{1}{2}$ came from? The original exponent of $-\frac{1}{2}$ was raised by 1, to give $-\frac{1}{2} + 1 = \frac{1}{2}$. The 2 comes from *dividing* by the new exponent of $\frac{1}{2}$.

Example 10.6 Find $\displaystyle\int \frac{1}{x}\, dx$.

Solution: Of course, $\frac{1}{x}$ can be written as x^{-1}. But if you try using the power rule and raising the power by 1, you will get $x^0 = 1$. As the derivative of the constant 1 equals 0, this clearly is not the right answer. If you try dividing by the new power, you get $\frac{1}{0}x^0$, which makes no sense. If you look again at the statement of Theorem 10.3, you will see mentioned this one exception. The power rule for integrals does not work when the original exponent equals -1. This is an antiderivative that you do not yet know how to find.

Note to students: The function whose derivative equals $\frac{1}{x}$ is the natural logarithm function, $\ln(x)$.

Using Theorem 5.3 backwards, you get

THEOREM 10.4. *Let c be a constant and $f(x)$ a function. Then*

1.

$$\int cf(x)\, dx = c \int f(x)\, dx + C$$

2. *The integral of a constant times a function equals the constant times the integral of the function.*

Proof: Suppose that $F(x)$ is an antiderivative of $f(x)$, so that $F' = f$. Then by Theorem 5.3,

$$(cF)' = cF' = cf$$

Example 10.7 Evaluate $\int 7x^2\, dx$.

Solution:

$$\int 7x^2\, dx = 7 \cdot \int x^2\, dx = 7 \cdot \frac{1}{3}x^3 + C = \frac{7}{3}x^3 + C$$

Example 10.8 Compute $\int \frac{1}{7t^4}\, dt$.

Solution:

$$
\begin{aligned}
\int \frac{1}{7t^4}\, dt &= \int \frac{1}{7} \cdot \frac{1}{t^4}\, dt \\
&= \frac{1}{7} \cdot \int t^{-4}\, dt \\
&= \frac{1}{7} \cdot \frac{1}{-3} t^{-3} + C \\
&= \frac{-1}{21t^3} + C
\end{aligned}
$$

THEOREM 10.5.

1. *(a)*

$$\int [f(x) + g(x)]\, dx = \int f(x)\, dx + \int g(x)\, dx$$

(b) *The integral of a sum equals the sum of the integrals.*

2. (a)
$$\int [f(x) - g(x)]\, dx = \int f(x)\, dx - \int g(x)\, dx$$

(b) *The integral of a difference equals the difference of the integrals.*

Proof: Of course the proof follows because an integral is an antiderivative, and because of Theorems 5.4 and 5.5.

With the above theorems, you can find the integral or antiderivative of any polynomial.

Example 10.9 Find $\displaystyle \int \left(2x^4 - 4x^2 + 7\right)\, dx.$

Solution:

$$
\begin{aligned}
\int \left(2x^4 - 4x^2 + 7\right)\, dx &= \int 2x^4\, dx - \int 4x^2\, dx + \int 7\, dx \\
&= 2\int x^4\, dx - 4\int x^2\, dx + 7\int 1\, dx \\
&= 2\cdot \frac{1}{5}x^5 - 4\cdot \frac{1}{3}x^3 + 7x + C \\
&= \frac{2}{5}x^5 - \frac{4}{3}x^3 + 7x + C
\end{aligned}
$$

Example 10.10 Find $\displaystyle \int \frac{x^4 + x}{x^3}\, dx.$

Solution: The integrand can be written as a sum of powers of x.

$$
\begin{aligned}
\int \frac{x^4 + x}{x^3}\, dx &= \int \left(\frac{x^4}{x^3} + \frac{x}{x^3}\right)\, dx \\
&= \int \left(x + x^{-2}\right)\, dx \\
&= \int x\, dx + \int x^{-2}\, dx \\
&= \frac{1}{2}x^2 + \frac{1}{-1}x^{-1} + C = \frac{1}{2}x^2 - \frac{1}{x} + C
\end{aligned}
$$

A Little More Notation

You now know how to find some antiderivatives, and you also know that evaluating a *definite* integral is a two-step process: find *any* antiderivative F, and then compute $F(b) - F(a)$. A notation has been devised to separate the two steps of this process so that you do not have to do everything at once.

DEFINITION 10.2. *For any function $F(x)$, the notation* $F(x)\Big|_a^b$ *means simply* $F(b) - F(a)$.

Example 10.11 Evaluate $x^2 \Big|_2^4$.

$\quad$ ***Solution:*** $x^2 \Big|_2^4 = 4^2 - 2^2 = 16 - 4 = 12$

When you have to evaluate a *definite* integral, first find any antiderivative and write it in the above notation, that is, followed by a vertical line and the limits of integration written in. *Then* compute the final answer.

Example 10.12 Evaluate $\displaystyle\int_2^4 3x^2\,dx$.

$\quad$ ***Solution:***

$$\int_2^4 3x^2\,dx = x^3 \Big|_2^4 = 4^3 - 2^3 = 64 - 8 = 56$$

Revisiting Problem 1 of Section 8.1 *for the last time*:

Example 10.13 A bowling ball has a velocity of $v(t) = 9 - 0.2t$ feet per second, t seconds after it has been let go. How far does the ball roll during the first two seconds after it has been let go?

$\quad$ ***Solution:*** You already know from Example 8.17 of Section 8.3 that the answer is given by the definite integral $\int_0^2 (9 - 0.2t)\,dt$. Thus the answer is

$$\int_0^2 (9 - 0.2t)\,dt = \left(9t - \frac{0.2}{2}t^2\right)\Big|_0^2$$
$$= (9t - 0.1t^2)\Big|_0^2$$
$$= [9\cdot 2 - 0.1\cdot(2^2)] - [9\cdot 0 - 0.1\cdot(0^2)]$$
$$= (18 - 0.4) - (0 - 0) = 17.6$$

A lot has gone into this "easy" way of solving the problem. But after putting in all of that preliminary effort, you have to admit that the solution has become easy.

Revisiting Problem 2 of Section 8.1 *for the last time:*

Example 10.14 Find the area of the region in the xy-plane bounded below by the x-axis, on the left by the vertical line $x = 1$, on the right by the vertical line $x = 5$, and above by the curve $y = x^2$.

Solution: By Example 8.18 of Section 8.3, the answer is given by the definite integral $\int_1^5 x^2\, dx$. Thus the answer is

$$
\begin{aligned}
\int_1^5 x^2\, dx &= \left. \frac{1}{3}x^3 \right|_1^5 \\
&= \frac{1}{3}\cdot 5^3 - \frac{1}{3}\cdot 1^3 \\
&= \frac{1}{3}\cdot 125 - \frac{1}{3}\cdot 1 \\
&= \frac{125-1}{3} = \frac{124}{3} = 41\frac{1}{3}
\end{aligned}
$$

10.5 ANTIDERIVATIVES INVOLVING THE CHAIN RULE

In this section, you will learn how to find the antiderivatives of some slightly more complicated functions. The antiderivatives of this section will involve using the chain rule of Section 5.3. As in the last section, the best way to see the patterns involved in antiderivatives is to know really well the rules and formulas for computing derivatives. If you have forgotten about the chain rule, go back now and review Section 5.3 and work all the examples in that section.

Suppose you want to find $\int (2x-3)^{12}\, dx$. You might at first think that this is just an example involving the power rule, and that the answer should just be $F(x) = \frac{1}{13}(2x-3)^{13}$, which is the result of raising the power by 1 and dividing by the new power. But if you check your answer by taking its derivative, and if you notice that $F(x)$ is a *composite* function, then you will remember that you have to use the *chain rule* to take the derivative of $F(x)$.

Let $y = \dfrac{1}{13}(2x-3)^{13}$. Then $y = \dfrac{1}{13}u^{13}$, where $u = 2x - 3$. According to the chain rule,

$$
\begin{aligned}
\frac{dy}{dx} &= \frac{dy}{du}\frac{du}{dx} \\
&= \frac{1}{13}\cdot 13u^{12}\cdot 2 \\
&= 2(2x-3)^{12}
\end{aligned}
$$

The first guess of $y = \dfrac{1}{13}(2x-3)^{13}$ did not give the right antiderivative for $(2x-3)^{12}$, but you can see that it wasn't off by too much. The derivative turned out to be twice as big as you wanted it to be, but you learned in the last section that if your initial attempt at finding an antiderivative is off by a multiplicative constant, then you can always fix things up to get the constant right. If your initial attempt gives a derivative twice as big as it should, then just divide by two. Following this logic,

$$
y = \frac{1}{2}\cdot\frac{1}{13}(2x-3)^{13} = \frac{1}{26}(2x-3)^{13}
$$

is an antiderivative of $(2x-3)^{12}$, as you can easily check.

The above explanation was meant to be fairly informal. The idea was to concentrate on the outside part of the expression $(2x-3)^{12}$ you wanted to integrate, namely the twelfth power, go with that, check your first attempt at an antiderivative, and adjust the answer. Since the derivative of the inside part of $(2x-3)^{12}$ is just 2, a constant, the first attempt could be adjusted to give a correct antiderivative.

If u stands for *any* function of x, and $y = u^n$ (so that y can also be considered a function of x), then the chain rule gives

$$
\begin{aligned}
\frac{dy}{dx} = \frac{d(u^n)}{dx} &= \frac{d(u^n)}{du}\frac{du}{dx} \\
&= nu^{n-1}\frac{du}{dx}
\end{aligned}
$$

Anything with the above pattern has an antiderivative of u^n. If your integrand has almost the above pattern, except that it differs from the above by a multiplicative constant, you can always adjust the multiplicative constant. Your antiderivative will be some constant, times a power of u one higher (with the exception of when the power of u is -1). It follows that you can deal with integrands of the form $u^n\dfrac{du}{dx}, n \neq -1$.

You can deal with integrands of the above form in two ways. The first is informal, and involves looking at the pattern of the integrand, guessing an antiderivative, taking the derivative of your guess, and adjusting by a multiplicative constant if that is all you are off by. That's the approach you went through above for finding $\int(2x-3)^{12}\,dx$.

There is a more formal method for attempting such integrals. The method is called **integration by substitution**. Integration by substitution is the mirror image of the approach to the chain rule that involves actually substituting a new letter "u" for the inside part of the composite. In integration by substitution, you try to substitute a new variable or letter (usually denoted by "u") for the inside part of a composite function, with the idea of trying to get your integral to look like $\int u^n\, du$, which would equal $\frac{1}{n+1}u^{n+1}$. The relationship between dx and du comes about from the formula

$$du = \frac{du}{dx}dx$$

This might sound a little confusing, but let's do the same integral, $\int (2x - 3)^{12}\, dx$, using integration by substitution. Just as with the similar approach to the chain rule for derivatives, try a substitution $u = 2x - 3$ of u for the inside part of the composite $(2x - 3)^{12}$. The term $(2x - 3)^{12}$ then just becomes u^{12}. So far, so good. What about dx? Well, If $u = 2x - 3$, then $\frac{du}{dx} = 2$, so that

$$du = \frac{du}{dx}dx = 2\, dx$$

In addition to $(2x-3)^{12}$ in the integrand, you would have liked the term $2\, dx$ in the integrand, because then you could have just replaced it with du. However, if $du = 2\, dx$, then $dx = \frac{1}{2}\, du$. Thus,

$$\begin{aligned}
\int (2x - 3)^{12}dx &= \int u^{12}\frac{1}{2}du \\
&= \frac{1}{2}\int u^{12}du \\
&= \frac{1}{2}\cdot\frac{1}{13}u^{13} \\
&= \frac{1}{26}u^{13} \\
&= \frac{1}{26}(2x - 3)^{13} + C
\end{aligned}$$

This is the same answer found earlier.

In the last line, of course, I both substituted the original variable and added the arbitrary constant. The advantage of integration by substitution is that in the computation above, the one integral computed was a completely standard integral, $\int u^{12}\, du$, using Theorem 10.3. The disadvantage of integration by substitution over the first approach is that it requires more writing and more work.

Unlike the formulas for derivatives, which can be applied in a completely mechanical fashion, finding antiderivatives sometimes requires insight into the pattern of the function you are trying to integrate. The best way to gain such insight is to be really solid in your understanding of the rules for derivatives, and especially of the chain rule, so that you can really see the kinds of patterns that come up when you differentiate certain kinds of functions. You will then be able to recognize these patterns when they arise in antiderivative or integral problems.

If the preceding problem of finding $\int (2x - 3)^{12} \, dx$ were changed to finding $\int (x^2 - 3)^{12} \, dx$, it would be impossible to do by the methods you have learned so far (unless you were willing to expand out the power). To see why, notice what happens if you try to use Theorem 10.3 to make a first attempt at finding an antiderivative. You would get $F(x) = \dfrac{1}{13}(x^2 - 3)^{13}$. Try to compute $F'(x)$ to see if it is really an antiderivative. If you compute the derivative of $F(x)$ correctly, using the chain rule and remembering to multiply by the derivative of the inside part, you will get

$$F'(x) = \frac{1}{13} \cdot 13(x^2 - 3)^{12} \cdot 2x = (x^2 - 3)^{12} \cdot 2x$$

You are off from what you want, $(x^2-3)^{12}$, not just by a multiplicative *constant*, for which you could always have adjusted, but by a factor of $2x$, which involves the *variable* as well. This cannot be adjusted for, and means that your approach is wrong.

On the other hand, if your problem were the more complicated looking problem of finding $\int x(x^2 - 3)^{12} \, dx$, you would be all set. Theorem 10.3 can be used. If you substitute $u = x^2 - 3$, then $\dfrac{du}{dx} = 2x$. Therefore,

$$du = \frac{du}{dx} dx = 2x \, dx$$

So, the expression $x \, dx$ in the integrand equals $\frac{1}{2} \, du$, and

$$\int x(x^2 - 3)^{12} \, dx = \int \frac{1}{2} u^{12} \, du = \frac{1}{2} \int u^{12} \, du$$

This last integral is completely standard.

Example 10.15 Find $\displaystyle\int x(x^2 - 3)^{12} \, dx$.

Solution:

METHOD 1: Try Theorem 10.3. Because the derivative of the inside part is $2x$, that should take care of the factor of x in the integrand, and you should just be off by a multiplicative constant for which you *can* adjust. So try $F(x) = \dfrac{1}{13}(x^2 - 3)^{13}$ as a first attempt at an antiderivative. Then

$$F'(x) = \frac{1}{13} \cdot 13(x^2 - 3)^{12} \cdot 2x$$
$$= 2x(x^2 - 3)^{12}$$

This is off by a factor of 2 from what you wanted, so adjust by dividing the first guess by 2, to get

$$\int x(x^2 - 3)^{12}\, dx = \frac{1}{2} \cdot \frac{1}{13}(x^2 - 3)^{13} = \frac{1}{26}(x^2 - 3)^{13} + C$$

METHOD 2: Try a substitution of u for the inside part of the composite, and hope that you will end up with an easier integral. If $u = x^2 - 3$, then $\dfrac{du}{dx} = 2x$, so that $du = 2x\, dx$, or $x\, dx = \dfrac{1}{2}\, du$. Making these substitutions, you get

$$\begin{aligned}
\int x(x^2 - 3)^{12}\, dx &= \int u^{12}\frac{1}{2}\, du \\
&= \int \frac{1}{2}u^{12}\, du \\
&= \frac{1}{2}\int u^{12}\, du \\
&= \frac{1}{2} \cdot \frac{1}{13}u^{13} \\
&= \frac{1}{26}(x^2 - 3)^{13} + C
\end{aligned}$$

If you try to substitute u for a function of x, then after you have made all the substitutions you should have replaced all the x's and dx inside the integral with u's and du. If you end up with both variables x and u, then your substitution will not work, and you should try a different substitution.

Example 10.16 Evaluate $\displaystyle\int 2x^2\sqrt{x^3 + 1}\, dx$.

Solution: As always, rewrite the $\sqrt{}$ sign in exponent form, so the problem is to evaluate $\int 2x^2(x^3 + 1)^{\frac{1}{2}}\, dx$.

METHOD 1: Try using Theorem 10.3 on $(x^3 + 1)^{\frac{1}{2}}$. Because the derivative of the inside part $x^3 + 1$ is $3x^2$, which takes care of the factor of x^2, it should work. Thus, a first attempt is trying $F(x) = \dfrac{2}{3}(x^3 + 1)^{\frac{3}{2}}$ as an antiderivative. Then

$$\begin{aligned}
F'(x) &= \frac{2}{3} \cdot \frac{3}{2}(x^3 + 1)^{\frac{1}{2}} \cdot 3x^2 \\
&= 3x^2(x^3 + 1)^{\frac{1}{2}}
\end{aligned}$$

You can see that you wanted a 2 in front of the expression $x^2(x^3 + 1)^{\frac{1}{2}}$ but you've got a 3 in front of it instead. Well, get rid of the 3 by dividing the original attempt $F(x) = \frac{2}{3}(x^3 + 1)^{\frac{3}{2}}$ by 3, and then put in the 2 by multiplying it by 2, and try

$$F(x) = \frac{1}{3} \cdot 2 \cdot \frac{2}{3}(x^3 + 1)^{\frac{3}{2}} = \frac{4}{9}(x^3 + 1)^{\frac{3}{2}} + C$$

Now, check the answer.

$$\begin{aligned}
\left(\frac{4}{9}(x^3 + 1)^{\frac{3}{2}}\right)' &= \frac{4}{9} \cdot \left((x^3 + 1)^{\frac{3}{2}}\right)' \\
&= \frac{4}{9} \cdot \frac{3}{2} \cdot (x^3 + 1)^{\frac{1}{2}} \cdot 3x^2 \\
&= 2x^2(x^3 + 1)^{\frac{1}{2}}
\end{aligned}$$

METHOD 2: Let $u = x^3 + 1$, so that $\dfrac{du}{dx} = 3x^2$, and $du = 3x^2\,dx$. The term $(x^3 + 1)^{\frac{1}{2}}$ in the integrand is replaced by $u^{\frac{1}{2}}$, while the term $x^2\,dx$ can be replaced by $\dfrac{1}{3}\,du$. Thus,

$$\begin{aligned}
\int 2x^2(x^3 + 1)^{\frac{1}{2}}\,dx &= \int 2u^{\frac{1}{2}} \cdot \frac{1}{3}\,du \\
&= \int \frac{2}{3}u^{\frac{1}{2}}\,du \\
&= \frac{2}{3}\int u^{\frac{1}{2}}\,du \\
&= \frac{2}{3} \cdot \frac{2}{3}u^{\frac{3}{2}} \\
&= \frac{4}{9}u^{\frac{3}{2}} = \frac{4}{9}(x^3 + 1)^{\frac{3}{2}} + C
\end{aligned}$$

Example 10.17 Evaluate $\displaystyle\int \frac{1}{\sqrt{3t + 1}}\,dt$.

Solution: Rewrite the integral as $\displaystyle\int \frac{1}{(3t + 1)^{\frac{1}{2}}}\,dt = \int (3t + 1)^{\frac{-1}{2}}\,dt$.

METHOD 1: Try $F(t) = 2(3t + 1)^{\frac{1}{2}}$, from the power rule for integrals. Then

$$\begin{aligned}
F'(t) &= 2 \cdot \frac{1}{2}(3t + 1)^{\frac{-1}{2}} \cdot 3 \\
&= 3(3t + 1)^{\frac{-1}{2}}
\end{aligned}$$

This is three times as big as you want it to be, so divide your first attempt by 3, and try $F(t) = \frac{2}{3} \cdot (3t + 1)^{\frac{1}{2}}$. For *this* $F(t)$,

$$
\begin{aligned}
F'(t) &= \frac{2}{3} \cdot \frac{1}{2} \cdot (3t + 1)^{\frac{-1}{2}} \cdot 3 \\
&= \frac{2}{3} \cdot \frac{1}{2} \cdot 3 \cdot (3t + 1)^{\frac{-1}{2}} \\
&= (3t + 1)^{\frac{-1}{2}}
\end{aligned}
$$

This is precisely what you wanted the derivative to equal.

METHOD 2: Let $u = 3t + 1$, so that $\frac{du}{dt} = 3$, $du = 3\,dt$, and $dt = \frac{1}{3}\,du$. When you make these substitutions,

$$
\begin{aligned}
\int (3t + 1)^{\frac{-1}{2}}\,dt &= \int u^{\frac{-1}{2}} \cdot \frac{1}{3}\,du \\
&= \frac{1}{3} \cdot \int u^{\frac{-1}{2}}\,du \\
&= \frac{1}{3} \cdot 2 \cdot u^{\frac{1}{2}} \\
&= \frac{2}{3} u^{\frac{1}{2}} = \frac{2}{3}(3t + 1)^{\frac{1}{2}} + C
\end{aligned}
$$

10.6 SUMMARY OF MAIN POINTS

- Integration and differentiation are inverse processes to each other. Doing one and then the other (sort of) gets you back to where you started.

- The theorem that expresses this relationship between integration and differentiation is called the **Fundamental Theorem of Calculus**:
  If $f(t)$ is continuous and a function $A(x)$ is defined by

$$
A(x) = \int_a^x f(t)\,dt
$$

  then $A(x)$ is differentiable, and $A'(x) = f(x)$

- It follows that integrals can be computed by finding **antiderivatives**. An antiderivative of $f(x)$ is a function $F(x)$ whose derivative equals $f(x)$, that is, $F'(x) = f(x)$.

- If a function $f(x)$ has an antiderivative, it has infinitely many, and they all differ by a constant. It follows that if $F(x)$ is one particular antiderivative of $f(x)$, then $F(x) + C$ is the **general** antiderivative of $f(x)$.

- It follows that to compute a **definite integral** $\int_a^b f(x)\,dx$, you just have to find *any* antiderivative $F(x)$, and then

$$\int_a^b f(x)\,dx = F(b) - F(a)$$

- To separate the work of finding an antiderivative $F(x)$ from the computation of $F(b) - F(a)$, the following notation has been devised:

$$F(x)\Big|_a^b = F(b) - F(a)$$

In evaluating a definite integral, first find an antiderivative and write it down, followed by the vertical line and the limits of integration. Then do the computation of $F(b)$, $F(a)$, and their difference.

- Some antiderivatives are easy to find by reversing the rules for derivatives. Examples of these follow:

  — The integral of 0 is a constant: $\int 0\,dx = C$

  — The integral of 1 is the variable of integration: $\int 1\,dx = x + C$

  — To find the integral of a power, add 1 to the exponent and divide by the new exponent. This works as long as the original exponent is not -1:

$$\int x^n\,dx = \frac{1}{n+1}x^{n+1} + C, \text{ for } n \neq -1$$

  — The integral of a constant times a function equals the constant times the integral of the function:

$$\int cf(x)\,dx = c\int f(x)\,dx$$

  — The integral of a sum is the sum of the integrals:

$$\int [f(x) + g(x)]\,dx = \int f(x)\,dx + \int g(x)\,dx$$

  — The integral of a difference is the difference of the integrals:

$$\int [f(x) - g(x)]\,dx = \int f(x)\,dx - \int g(x)\,dx$$

- Some integrals must be computed by using the chain rule in reverse. These can be done more informally, if you see the patterns that are involved, or more formally by the technique of integration by substitution. The integrals of this type that you have learned how to do in this book are of the form, more or less, $\int u^n \dfrac{du}{dx}\,dx$, where u is some function of x. Then

$$\int u^n \frac{du}{dx}\,dx = \frac{1}{n+1}u^{n+1} + C$$

This is because

$$
\begin{aligned}
\frac{d}{dx}\left(\frac{1}{n+1}u^{n+1}\right) &= \frac{d}{du}\left(\frac{1}{n+1}u^{n+1}\right)\frac{du}{dx} \\
&= \frac{1}{n+1}\cdot(n+1)u^n\frac{du}{dx} \\
&= u^n\frac{du}{dx}
\end{aligned}
$$

10.7 EXERCISES

For Exercises 1 and 2, first evaluate $F(x)$, and then compute the derivative $F'(x)$ of $F(x)$. (Reread Part I of the Fundamental Theorem of Calculus, in Section 10.3.)

1. $F(x) = \displaystyle\int_2^x 3t^2\,dt$

2. $F(x) = \displaystyle\int_1^x (\sqrt{t}+1)\,dt$

For Exercises 3 and 4, use the Fundamental Theorem of Calculus to find $F'(x)$.

3. $F(x) = \displaystyle\int_4^x \frac{1}{t}\,dt$

4. $F(x) = \displaystyle\int_0^x (t^3-1)^{11}\,dt$

For Exercises 5–15, find the indefinite integrals, or general antiderivatives.

5. $\displaystyle\int (2x^6 - 6x^9)\,dx$

6. $\displaystyle\int 3\sqrt{2t-1}\,dt$

7. $\displaystyle\int \frac{1}{(6x+4)^2}\,dx$

8. $\displaystyle\int \frac{x^4 + x^3 + 1}{x^2}\,dx$

9. $\displaystyle\int (2s^3 - 3s^2 + 4s - 7)\,ds$

10. $\displaystyle\int \frac{1}{\sqrt{y-4}}\, dy$

11. $\displaystyle\int \frac{4}{2x^{\frac{5}{2}}}\, dx$

12. $\displaystyle\int 3x^3\sqrt{2x^4 + 1}\, dx$

13. $\displaystyle\int \frac{t}{\sqrt{t^2 + 4}}\, dt$

14. $\displaystyle\int \frac{s^2}{(s^3 - 3)^7}\, ds$

15. $\displaystyle\int \frac{(\sqrt{x} + 1)^5}{\sqrt{x}}\, dx$

In Exercises 16–20, evaluate the definite integrals.

16. $\displaystyle\int_0^3 (2t^2 - 3t)\, dt$

17. $\displaystyle\int_2^4 \frac{3}{4x^2}\, dx$

18. $\displaystyle\int_{-2}^3 x(x^2 + 1)^4\, dx$

19. $\displaystyle\int_{-3}^{-1} (x^2 - 4x)\, dx$

20. $\displaystyle\int_{-1}^1 (x - 1)\, dx$

11
GEOMETRIC APPLICATIONS OF THE INTEGRAL

You have already seen, in the previous three chapters, how integrals can be used to compute the area of the region in the xy-plane under the graph of the positive function $f(x) = x^2$ and over the x-axis, between the vertical lines $x = 1$ and $x = 5$. In fact, integrals can be used to compute areas of much more general types of regions, and they can be used also to compute volumes, lengths of curves, areas of surfaces in three-dimensional space, and much more. It is only because of the *definition* of the integral as a limit of Riemann or approximate sums that you can see why geometry problems, such as finding areas of regions, can be solved by setting up an integral. After seeing that you can solve a problem by setting up an integral, you can then use the Fundamental Theorem of Calculus to evaluate the integral by means of antiderivatives.

11.1 HORIZONTAL VS. VERTICAL, *X* VS. *Y*

Setting up the correct integral to solve a geometry problem is not really that hard, but there are certain simple yet fundamental points you must remain clear about. Just about every geometry problem involves determining the lengths of some horizontal or vertical lines. You must always be completely clear about the difference between:

- left and right;
- bottom and top;
- x and y;
- horizontal and vertical.

The remainder of this section explains what this means.

Let's get started. How could you describe precisely the difference between the two lines in bold in Figure 11.1?

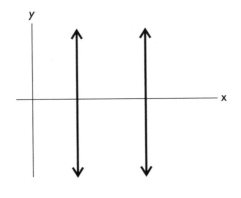

Figure 11.1

A good answer would be "they have different x-coordinates." A vertical line is specified by its x-coordinate. That means that all points on a vertical line have the *same x-coordinate*, while the y-coordinates of different points on the same vertical line differ. For example, $x = 2$ is the formula for the vertical line in the xy-plane consisting of all points with coordinates $(2, y)$. Every point on the line has an x-coordinate of 2, but you can't say anything in particular about the y-coordinates of points on the line $x = 2$, because the y-coordinates could be anything.

Thus the difference between the two vertical lines has to be described in terms of their different x-coordinates. One might be the line $x = 2$ while the other might be the line $x = 5$. It follows that if you have a whole bunch of vertical lines or line segments in a picture, then the different line segments must be distinguished from each other by their different x-coordinates. Thus any difference between one vertical line and the next must be described in terms of x, and when you have a picture like that in Figure 11.2, x must be the variable (more technically, the variable of integration).

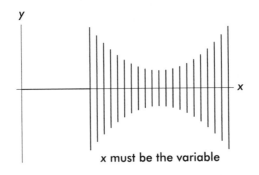

x must be the variable

Figure 11.2

How could you describe precisely the difference between the two lines in bold in Figure 11.3?

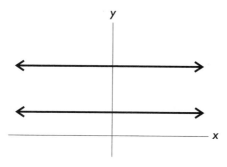

Figure 11.3

Well, different horizontal lines have different y-coordinates. A horizontal line is specified by its y-coordinate, because all points on the line have precisely the *same* y-coordinate, while any number can be an x-coordinate for some point on the line. Thus $y = 2$ is the equation for the set of all points in the plane whose y-coordinate is 2, and whose x-coordinate could be anything. Therefore the difference between the two horizontal lines has to be described in terms of their different y-coordinates. For example, the bottom line might be $y = 1$ while the top line might be $y = 3$.

It follows that if you have a whole bunch of horizontal lines or line segments in a picture (see Figure 11.4), then the different line segments must be distinguished from each other by their different y-coordinates. Thus the difference between the different horizontal line segments in your picture must be described in terms of y, and y must be the descriptive variable (that is, the variable of integration).

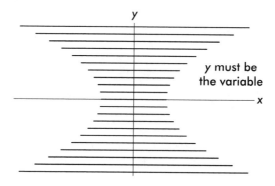

Figure 11.4

All you have to know beyond this is how to compute the length of a horizonal or vertical line segment. Suppose the bottom point of the vertical line segment of Figure 11.5 has coordinates (2, 3). What are plausible candidates for the coordinates of the top point on the line?

Figure 11.5

Well, lots of different answers are plausible, but the first coordinate has to be 2 (because all points on a vertical line have the same x-coordinate), and the second coordinate has to be larger than 3 (because in the y-direction, the convention is that larger means higher). Thus, (2, 5) or (2, 7) or (2, 8) are all plausible answers. To be specific, suppose the coordinates of the top point are (2, 7), as shown in Figure 11.6.

Figure 11.6

How long is the segment? Write out for yourself in words what you did to get the answer.

If you wrote "I subtracted 3 from 7," write the answer out in words in a way that somebody else could use to find the length of a different vertical segment. A good answer would be the following:

> The length of a vertical line segment is computed by taking the y-coordinate of the point on the top of the vertical segment and subtracting the y-coordinate of the point on the bottom.

More briefly,

The length of a vertical line = (y on top) − (y on bottom).

This holds for all vertical segments, whether both the top and the bottom are above the x-axis, or one is above and the other below, or both are below (see Figure 11.7).

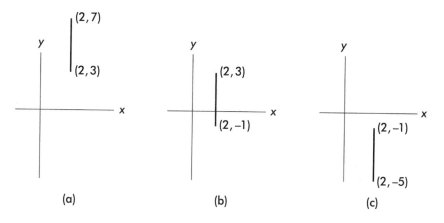

(a)   (b)   (c)

Figure 11.7

In (a), the length equals $7 - 3 = 4$, in (b) the length equals $3 - (-1) = 3 + 1 = 4$, and in (c) the length equals $-1 - (-5) = -1 + 5 = 4$.

If you compute the y-coordinate of the point on the top minus the y-coordinate of the point on the bottom, you will always get the correct answer for the length of a vertical line segment.

Consider the horizontal line segment of Figure 11.8, whose right-hand endpoint has coordinates $(1, 3)$.

 (1,3)

Figure 11.8

What are plausible candidates for the coordinates of the left-hand endpoint? Again, lots of answers are plausible, but the second coordinate, the y-coordinate, must be 3 (because all points on the same horizontal line have the same y-coordinate), and the first coordinate, the x-coordinate, must be smaller than 1 (because in the x-direction, the convention is that larger means to the right, so smaller means to the left). Thus ($\frac{1}{2}$, 3) or (0, 3) or (−3, 3) or (−4, 3) are all plausible answers.

To be specific, suppose the left-hand endpoint has coordinates $(-3, 3)$, as in Figure 11.9.

$$(-3,3) \text{————————} (1,3)$$

Figure 11.9

What is the length of the segment? Write down in words how you computed its length. Make sure your description is general enough to serve as a recipe for someone computing the length of any other horizontal line segment.
A good answer would be the following:

> The length of a horizontal line segment is computed by taking the x-coordinate of the point on the right of the horizontal segment and subtracting the x-coordinate of the point on the left.

More briefly,

> The length of a horizontal line = (x on right) − (x on left).

This holds for all horizontal segments, whether both the left-hand and the right-hand endpoints are to the right of the y-axis, or one is to the right and the other to the left of the y-axis, or both are to the left of the y-axis (see Figure 11.10).

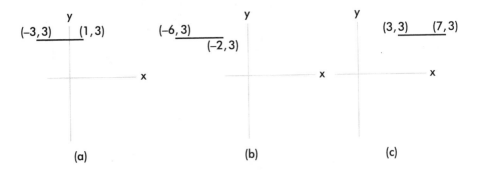

Figure 11.10

In (a) the length of the horizontal segment equals $1 - (-3) = 1 + 3 = 4$, in (b) the length is $(-2) - (-6) = -2 + 6 = 4$, and in (c) it is $7 - 3 = 4$.
If you compute the x-coordinate of the point on the right minus the x-coordinate of the point on the left, you will always get the correct answer for the length of a horizontal line segment.

Let's look again at the picture in Figure 11.11, which you've seen before in this section.

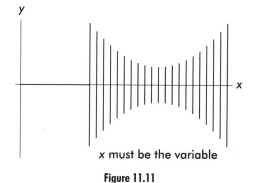

x must be the variable

Figure 11.11

You've seen that you have to use x as a variable to distinguish between the different vertical line segments. Suppose you wanted to compute the lengths of *all* the vertical line segments. As there are a lot of different segments, and they are not all of the same length, it is not a question of computing a number. Rather, it is a matter of writing down the length of the segment at position x in terms of x, that is, of getting a formula in terms of x for the length of the vertical segment at x. You've also seen that the length of a vertical segment is given by computing

(y-coordinate of the point on the top) − (y-coordinate of the point on the bottom)

That is, you need to know the y-coordinates of the points on the top and the y-coordinates of the points on the bottom, in terms of x. Knowing the y-coordinates of a bunch of points in terms of x just means knowing that the points are on the graph of some function $y = f(x)$. Thus you need to know that the top points are on the graph of some function $y = f(x)$ and the bottom points are on the graph of (some other) function $y = g(x)$ (see Figure 11.12).

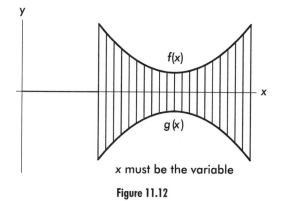

x must be the variable

Figure 11.12

Then for the vertical line segment at position x, its length is given by computing

(y-coordinate of the point on the top) − (the y-coordinate of the point on the bottom) $= f(x) - g(x)$

If you want to get a hint of where all of this is going, compare Figure 11.12 with that of Figures 8.4, 8.5, and 8.6 in Chapter 8.

Now let's look at the picture in Figure 11.13, which you've also seen before in this section.

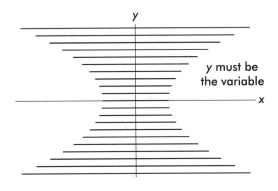

y must be the variable

Figure 11.13

You've seen that you have to use y as a variable to distinguish between the different horizontal line segments. Suppose you wanted to compute the lengths of *all* of the horizontal line segments. As there are a lot of different segments, and they are not all of the same length, it is not a question of computing a number. Rather, it is a matter of writing down the length of the segment at position y in terms of y, that is, of getting a formula in terms of y for the length of the horizontal segment at y. You've also seen that the length of a horizontal segment is given by computing

(x-coordinate of the point on the right) − (x-coordinate of the point on the left)

That is, you need to know the x-coordinates of the points on the right, and the x-coordinates of the points on the left, in terms of y. Knowing the x-coordinates of a bunch of points in terms of y just means knowing that the points are on the graph of some function $x = k(y)$. Thus you need to know that the points on the right are on the graph of some function $x = k(y)$, and the points on the left are on the graph of (some other) function $x = h(y)$ (see Figure 11.14).

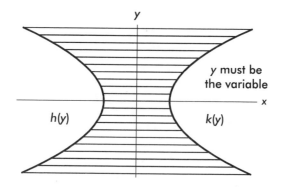

Figure 11.14

Then for the horizontal line segment at position y, its length is given by computing

(x-coordinate of the point on the right) $-$ (x-coordinate of the point on the left) $= k(y) - h(y)$

What could this possibly be good for? Well, look at the picture. The striped region is divided into a bunch of horizontal strips, which are almost rectangles. If you pretend they are rectangles, you can estimate the area of each strip by multiplying the height times the width. The height is just a little bit of the y-axis, which you can write as Δy. The width is the length of a horizontal segment. If you pick a sample point p in a small subinterval of the y-axis, then $k(p) - h(p)$ would be a good estimate for the width of the segment. Thus $(k(p) - h(p)) \cdot \Delta y$ would be a good estimate for the area of a little horizontal strip, and $\Sigma_{i=1}^{n}(k(p_i) - h(p_i))\Delta y$ would be a good estimate for the area of the striped region. It follows that the way to compute the *exact* area of the region is to find the following limit:

$$\lim_{n \to \infty} \sum_{i=1}^{n}(k(p_i) - h(p_i))\Delta y = \int [k(y) - h(y)]\, dy$$

11.2 AREA

Integrals can be used to compute the areas of very general regions in the plane. The basic idea, however, is similar to that used in computing the area bounded above by $y = x^2$, below by the x-axis, on the left by the line $x = 1$, and on the right by the line $x = 5$, which you studied in Chapters 8, 9, and 10.

1. Break the area up into thin horizontal or vertical strips.

2. Remember from Section 11.1 that if you break the area up into thin vertical strips, then the integral must be set up in terms of x, while if you break the area up into thin horizontal strips, the integral must be set up in terms of y.

3. Estimate the area of each strip by assuming the strip is a rectangle.

4. Add up the estimates for the area of each strip to get a Riemann or approximate sum, which will be an *estimate* for the whole area.

5. In the limit, letting the strips get thinner and thinner, you get the exact answer.

6. The limit of the approximate sums is an integral (by Definition 8.2 of Section 8.3), which by the Fundamental Theorem of Calculus (Chapter 10) can be computed by finding an antiderivative.

In all problems, you should draw graphs of the given curves, and get a picture of the region whose area you are asked to compute.

Example 11.1 Find the area of the region bounded above by $y = x^3$, below by $y = x^2$, on the left by $x = 2$, and on the right by $x = 3$.

Solution: The region is pictured in Figure 11.15.

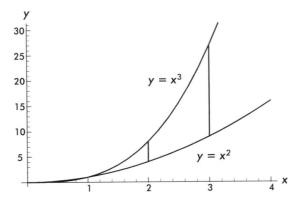

Figure 11.15

Unlike Problem 2 of Section 8.1, where the bottom of the region was part of the x-axis, the bottom here is another curve. But that hardly matters. If the top and bottom boundaries of the region were given by constant functions, you would have a rectangle, and the area would just be the height times the width. Here the problem is more complicated because the top and bottom are given by varying functions. However, since the functions vary continuously, they vary very little over small subintervals of the x-axis. It follows, therefore, that if you break the x-axis up into small intervals, and break the region up accordingly, as pictured in Figure 11.16, then each little strip is almost a rectangle. Furthermore, it follows from Section 11.1 and Figure 11.2 that everything in the integral must be set up in terms of x.

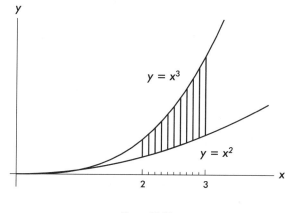

Figure 11.16

Focus on one little strip in the picture. It is very thin. Its width corresponds to just a little bit of the x-axis. The notation for the width of a small subinterval of the x-axis is Δx, or, if you want to think of the little strip as the ith little strip (say, out of n strips altogether), then you could denote the width by Δx_i (see Equations 8.3 or 8.4 of Section 8.2).

What about the height of the rectangle? If you let p (or p_i) be any sample point in the small interval (see Figure 11.17), then the part of the vertical line $x = p$ (or $x = p_i$) that is inside the region is a good representative sample for the height of the strip.

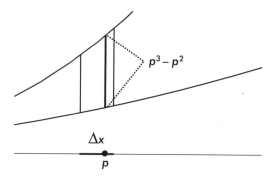

Figure 11.17

In the last section, you learned how to measure the length of a vertical line segment. It is given by

(y-coordinate of the point on the top) − (y-coordinate of the point on the bottom)

Since the top and bottom points lie on curves whose formulas you were given, you know formulas for these y-coordinates as functions of x. That's exactly what you want, because if you compare Figure 11.16 with Figure 11.2 of Section 11.1, you'll see that x should be the variable, that is, that everything should be computed in terms of x. Now when $x = p_i$, the y-coordinate on top is $y = p_i^3$, and the y-coordinate of the point on the bottom is $y = p_i^2$. Thus a representative height of the strip is $p_i^3 - p_i^2$. Accordingly, since the strip is approximately a rectangle, its area is approximated simply by multiplying the width times the height, as you would do for a rectangle. Thus, the area of the ith strip is approximately $\left(p_i^3 - p_i^2\right) \Delta x_i$. Adding up the estimates for the areas of all of the thin strips to get an estimate for the area of the whole region, you get that the area of the whole region is approximately

$$\sum_{i=1}^{n} \left(p_i^3 - p_i^2\right) \Delta x_i$$

Your estimates get better and better as you have more and more strips getting thinner and thinner. Recall that this is because on a smaller subinterval the continuous functions $y = x^3$ and $y = x^2$ vary less and less, so they are closer to being approximately constant. Thus a thinner vertical strip of area is closer to being a rectangle. In any case, you get in the limit the exact area,

$$\lim_{n \to \infty} \sum_{i=1}^{n} \left(p_i^3 - p_i^2\right) \Delta x_i$$

Now remember that the interval you are dividing into small subintervals is the interval $[2, 3]$, so by Section 8.3 the shorthand notation for the above limit is the integral

$$\int_2^3 \left(x^3 - x^2\right) dx$$

Recall from Section 8.3 that "in the limit" the summation sign becomes the integral sign, the limits of integration tell you which interval on the axis you are dividing up into lots of little pieces, the Δx becomes dx, and the p for a representative sample point reverts back to the common letter used for the variable, here x.

Finally, as the last step, recall from Chapter 10 that to evaluate the integral, you need only apply the Fundamental Theorem of Calculus and find an antiderivative. Thus

$$
\begin{aligned}
\int_2^3 (x^3 - x^2)\, dx &= \left(\frac{1}{4}x^4 - \frac{1}{3}x^3\right)\Big|_2^3 \\
&= \left(\frac{1}{4} \cdot 3^4 - \frac{1}{3} \cdot 3^3\right) - \left(\frac{1}{4} \cdot 2^4 - \frac{1}{3} \cdot 2^3\right) \\
&= \left(\frac{81}{4} - \frac{27}{3}\right) - \left(\frac{16}{4} - \frac{8}{3}\right) \\
&= 20\frac{1}{4} - 9 - 4 + 2\frac{2}{3} \\
&= 9\frac{11}{12}
\end{aligned}
$$

You've probably noticed that when you pass to a limit and go from summation notation to integral notation, the subscript i that keeps track of which little interval you are looking at disappears. You've probably also noticed that the new letter p that was introduced for a particular sample point changes back, in integral notation, to the common variable name x. To make life a little simpler in the future, you can just ignore the i subscript, and the introduction of the new letter p. Use x instead of p_i, but remember that x is playing two roles—it is sometimes the common name of the variable on the horizontal axis, and it is sometimes a particular, but arbitrary, value of that variable.

I did the above example in a lot of detail, partly to provide a review of some of the fundamental concepts, notation, and terminology that you have been studying for the past several chapters. The remaining examples in this section will be done in less detail, and will serve to illustrate some of the extra complications that can arise when doing area problems.

Example 11.2 Find the area of the region bounded on top by the curve $y = x^2 - 1$, on the bottom by the curve $y = -x^2 - 1$, on the left by the vertical line $x = 0$, and on the right by the vertical line $x = 3$.

Solution: The region is pictured in Figure 11.18. You don't have to be a Rembrandt, but you do have to sketch the rough features of the curves accurately.

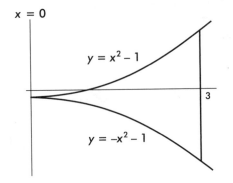

Figure 11.18

Breaking the region up into thin vertical strips, you see that x should be the variable, and that you can estimate the area of each little strip by multiplying the width Δx by the height (see Figure 11.19).

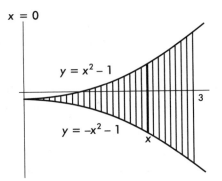

Figure 11.19

The height at a sample point x is given by

(y-coordinate on the top of the vertical segment) − (y-coordinate on the bottom)

Namely,

$$
\begin{aligned}
\text{height} &= (y \text{ on top}) - (y \text{ on bottom}) \\
&= (x^2 - 1) - (-x^2 - 1) = x^2 - 1 + x^2 + 1 = 2x^2
\end{aligned}
$$

The point of this example is that you do not have to worry about the fact that one of the curves is below the x-axis, and that the other curve is sometimes above the x-axis and sometimes below. Regardless of where the top and bottom points of the vertical segment are, simply computing (y on top) − (y on bottom) gives the right answer for its height.

Thus a thin strip containing the vertical segment at x has area approximately equal to $(2x^2)\,\Delta x$. If you sum up the areas of all these strips, you get an estimate for the area of the whole region. Passing to the limit, the sum becomes an integral, the Δx becomes dx, and the limits of integration are 0 and 3, because it is the interval $[0,3]$ on the x-axis that you are breaking up into little pieces. Thus the exact area equals

$$\int_0^3 2x^2\,dx \;=\; \frac{2}{3}x^3\Big|_0^3$$

$$=\; \left(\frac{2}{3}\cdot 3^3\right) - \left(\frac{2}{3}\cdot 0\right) = 18$$

In both of the previous examples, the left boundary and the right boundary of the region whose area you were asked to compute were stated explicitly. In some problems this is not done. That is because the curves cross at several points, and the region is meant to be the bounded region between the crossing points. Part of the work you are supposed to do, in such a problem, is figure out what the coordinates of the crossing points are, so that you can set up the limits of integration. This usually requires a little algebra.

A Review of Some Graphing and Algebra Techniques

Recall from Section 1.3 that if $y = f(x)$ is an equation, then a point in the plane with coordinates (a, b) is on the graph of the equation precisely when the y-coordinate b equals $f(a)$. If $y = f(x)$ and $y = g(x)$ are two equations, then they cross precisely when there is a point (a, b) that is on both graphs. This means that both $b = f(a)$ and $b = g(a)$. Thus there is a point $x = a$ for which $f(a) = g(a)$. Usually the way to find when the two curves cross is to find this point. That means setting the formulas for f and for g equal, and solving for x. Each solution for such an x gives a value $x = a$ at which $f(a) = g(a)$. With b equal to the common value $f(a) = g(a)$, (a, b) will be a point on both curves, and hence a crossing point.

After setting both formulas equal, and writing $f(x) = g(x)$, the standard way of solving for x is to move all terms to one side of the equation, so that the other side is 0, and then to simplify and factor the first side as much as possible. The first side is 0 exactly when any one of its factors is 0. When you do this, be careful not to throw away any factors, no matter how simple they are. Each factor can give rise to a crossing point of the two curves.

When you have all the crossing points of the curves, you will have to figure out, between each pair of consecutive crossing points, which curve is on the top and which is on the bottom. The easiest general way to do this is by the sample point method. For example, if two curves $f(x)$ and $g(x)$ cross at $x = 2$ and at $x = 4$, and nowhere in between, that means that on the whole interval $[2, 4]$, one of them is on the top and the other on the bottom. To see which, pick any sample point between 2 and 4, such as $x = 3$, and evaluate both functions there. If, for example, $g(3) > f(3)$, then since the values of the curves are plotted on the y-axis and a larger y-value means a higher point, the curve $y = g(x)$ is above the curve $y = f(x)$ at the point $x = 3$, and thus on the whole interval $[2, 4]$. If two adjacent crossing points are at $x = 2$ and $x = 3$, you will have to pick a fractional point in between, like $x = 2.5 = \dfrac{5}{2}$, as a sample point.

End of Review

In all area problems, you must be careful to sketch the curves accurately enough so that you can be sure you have all crossing points correctly, and so that you know for sure, between crossing points, which curve is on the top and which is on the bottom. If the curves do not cross, you still must know which one is on the top and which is on the bottom. To simplify things, I didn't mention any of this in the first two examples. There the curves did not cross (at least not in the interval the examples were looking at), and the top and bottom were drawn correctly. You really should have first checked whether there were any crossing points in the interval being looked at, and you should have also checked which curve was on the top and which was on the bottom.

Example 11.3 Find the area of the region bounded by the curves $y = (x - 1)^2$ and $y = (x - 1)^3$.

> *Solution:* Since no left and right boundary lines are mentioned, you can assume that the curves cross in at least two points, and that it is the region between the crossing points whose area you are supposed to compute. To find the crossing points, set the formulas equal, move everything to one side, simplify and factor, and solve:
>
> $$\begin{aligned} (x - 1)^2 &= (x - 1)^3 \\ (x - 1)^2 - (x - 1)^3 &= 0 \\ (x - 1)^2[1 - (x - 1)] = (x - 1)^2(2 - x) &= 0 \end{aligned}$$
>
> This gives $(x - 1)^2 = 0$ so that $x = 1$, and $2 - x = 0$ so that $x = 2$. At a point between $x = 1$ and $x = 2$, say $x = \dfrac{3}{2}$, you have

$$\left(\frac{3}{2}-1\right)^2 = \left(\frac{1}{2}\right)^2 = \frac{1}{4}, \text{ while } \left(\frac{3}{2}-1\right)^3 = \left(\frac{1}{2}\right)^3 = \frac{1}{8} < \frac{1}{4}.$$

Thus on the interval $[1, 2]$, the graph of $y = (x-1)^3$ is below that of $y = (x-1)^2$. The picture is shown in Figure 11.20.

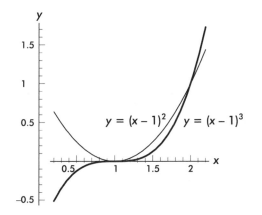

Figure 11.20

Divide the interval $[1, 2]$, and the corresponding region between the curves, into lots of little pieces, as shown in Figure 11.21.

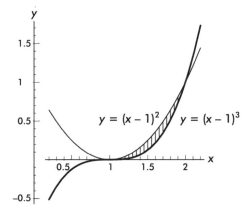

Figure 11.21

It follows that x is the variable of integration, and each little vertical strip is almost a rectangle, with width Δx and height at a general position x given by (y on top) − (y on bottom), namely, by $(x-1)^2 - (x-1)^3$. The area of the region is approximated by summing up, over the interval $[1, 2]$, the approximate areas $[(x-1)^2 - (x-1)^3]\Delta x$ of the different strips. The exact area is

$$\lim_{n\to\infty} \sum_{i=1}^{n} [(x_i - 1)^2 - (x_i - 1)^3]\Delta x_i$$

Or, rewriting this limit in integral notation,

$$
\begin{aligned}
\int_1^2 [(x-1)^2 - (x-1)^3]\,dx &= \left[\frac{1}{3}(x-1)^3 - \frac{1}{4}(x-1)^4\right]\Big|_1^2 \\
&= \left[\frac{1}{3}(2-1)^3 - \frac{1}{4}(2-1)^4\right] \\
&\quad - \left[\frac{1}{3}(1-1)^3 - \frac{1}{4}(1-1)^4\right] \\
&= \left[\frac{1}{3} - \frac{1}{4}\right] - [0 - 0] = \frac{1}{12}
\end{aligned}
$$

Example 11.4 Find the area of the region bounded by the curves $y = (x-1)^2$ and $y = (x-1)^3$, between $x = \dfrac{3}{2}$ on the left and $x = \dfrac{5}{2}$ on the right.

Solution: You've already done most of the work of drawing the curves and figuring out where they cross. The new thing here is that on part of the interval one of the two curves is on the top, and on the other part of the interval the other curve is on the top (see Figure 11.22).

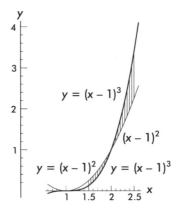

Figure 11.22

If you just set up an integral $\int_{3/2}^{5/2} [(x-1)^2 - (x-1)^3] \, dx$, then on the subinterval $\left[2, \frac{5}{2}\right]$ of the interval $\left[\frac{3}{2}, \frac{5}{2}\right]$ you'll be estimating the height of a thin strip by computing

(y-coordinate of the bottom point) − (y-coordinate of the top point).

This is of course negative, and is exactly the reverse of what you should be doing. You will be computing area, but it will be counted negatively. This negatively counted area will then cancel with the positively counted area for that part of the interval where $y = (x-1)^2$ is really on top. Your answer will really be a *difference* of two areas, the area between the graphs and above the interval $\left[\frac{3}{2}, 2\right]$ and the area between the graphs and above the interval $\left[2, \frac{5}{2}\right]$. Of course, if you reverse the functions and compute $\int_{3/2}^{5/2} [(x-1)^3 - (x-1)^2] \, dx$, you will be making a similar mistake.

The only solution is to break the interval up into two parts, $\left[\frac{3}{2}, 2\right]$ and $\left[2, \frac{5}{2}\right]$, compute each of the two areas separately, and add them up. Now in the previous example you've seen that on $\left[\frac{3}{2}, 2\right]$ the function $y = (x-1)^2$ is on top, while on the interval $\left[2, \frac{5}{2}\right]$ you can check, by evaluating both functions at $x = \frac{5}{2}$, that the function $y = (x-1)^3$ is on top. Thus the area is given by

$$\int_{3/2}^{2} [(x-1)^2 - (x-1)^3] \, dx + \int_{2}^{5/2} [(x-1)^3 - (x-1)^2] \, dx$$

$$= \left[\frac{1}{3}(x-1)^3 - \frac{1}{4}(x-1)^4 \right]\Big|_{3/2}^{2} + \left[\frac{1}{4}(x-1)^4 - \frac{1}{3}(x-1)^3 \right]\Big|_{2}^{5/2}$$

I'll leave the arithmetic up to you. Just remember, each of the two integrals should evaluate to a positive number.

Example 11.5 Find the area of the region between the curve $y = x^3 - x$ and the x-axis.

Solution: This problem is like the previous one. The x-axis is the same as the line with equation $y = 0$, so that to see where the curves cross, you set the equations equal, see that moving everything to one side is unnecessary, factor, and solve. Thus,

$$x^3 - x = x(x^2 - 1) = x(x - 1)(x + 1) = 0$$

so that the curve $y = x^3 - x$ crosses the x-axis at $x = -1, 0$, and 1. The graph, with the region divided up into thin vertical strips, is shown in Figure 11.23.

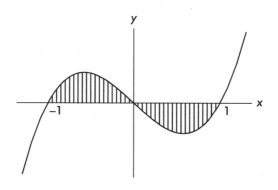

Figure 11.23

The area is

$$\int_{-1}^{0} (x^3 - x)\, dx + \int_{0}^{1} (-x^3 + x)\, dx$$

This is because on the interval $[-1, 0]$,

$$(y \text{ on top}) - (y \text{ on bottom}) = (x^3 - x) - 0 = x^3 - x$$

while on the interval $[0, 1]$,

$$(y \text{ on top}) - (y \text{ on bottom}) = 0 - (x^3 - x) = -x^3 + x$$

Do the computations yourself, and notice that each piece gives the same area. You can see from the graph that this would happen. This symmetry often gives an easier way to solve problems. You just have to compute the area of one of the regions, which you can do with one integral, and multiply by 2.

Example 11.6 Find the area of the region bounded by the curves $x = y^2$ and $y = x - 2$.

Solution: The graph of $x = y^2$ is like the graph of $y = x^2$, but with x and y reversed, so that instead of being a parabola with vertex at the origin, opening up, it is a parabola with vertex at the origin, opening to the right. Since no limits of integration are explicitly mentioned, you must figure that the curves meet at two points, and these two crossing points serve to mark the region under consideration. Now setting the y formulas equal and solving for x would be a nuisance here, since it would require starting with the formula $x = y^2$ and taking the square root of both sides. You would then have an equation in x to solve, involving $\sqrt{x}$. It is easier to set the x-formulas equal, and solve for y. Now the x-formula for the first curve is given explicitly as $x = y^2$. The second curve is given as $y = x - 2$, and getting the x formula for it just means solving for x in terms of y. This is easy enough. If $y = x-2$, then $x = y + 2$. Thus, setting the x-formulas equal, you get

$$
\begin{aligned}
y^2 &= y + 2 \\
y^2 - y - 2 &= 0 \\
(y - 2)(y + 1) &= 0
\end{aligned}
$$

So, $y = -1$, or $y = 2$. When $y = -1$, $x = y^2 = 1$, and when $y = 2$, $x = y^2 = 4$. Thus the two curves cross at the points $(1, -1)$ and $(4, 2)$. The curves are pictured in Figure 11.24.

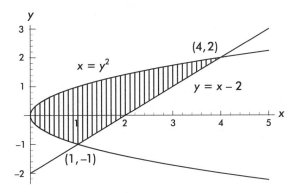

Figure 11.24

Note that if you draw vertical strips as above and use x as the variable, when it comes time to computing the height of a sample rectangle by $(y$ on top$) - (y$ on bottom$)$, there is a problem. The y on top is always on the part of the curve $x = y^2$ above the x-axis, so that $y = \sqrt{x}$, but the y on bottom lies on two different curves. Between $x = 0$ and $x = 1$, the y on bottom lies on the part of the curve $x = y^2$ below the x-axis, so that y is negative, and $y = -\sqrt{x}$, but between $x = 1$ and $x = 4$, the y on bottom lies on the line $y = x - 2$. In computing this integral the usual way, there is no option but to handle this problem by breaking the integral up into two parts, to reflect the fact that the bottoms of the vertical line segments lie on two different curves. Thus the area would be given by

$$\int_0^1 [\sqrt{x} - (-\sqrt{x})]\, dx + \int_1^4 [\sqrt{x} - (x - 2)]\, dx$$

$$= \int_0^1 2x^{\frac{1}{2}}\, dx + \int_1^4 \left(x^{\frac{1}{2}} - x + 2\right) dx$$

$$= \left.\left(2 \cdot \frac{2}{3} x^{\frac{3}{2}}\right)\right|_0^1 + \left.\left(\frac{2}{3} x^{\frac{3}{2}} - \frac{1}{2} x^2 + 2x\right)\right|_1^4$$

$$= \left(\frac{4}{3} \cdot 1^{\frac{3}{2}} - \frac{4}{3} \cdot 0^{\frac{3}{2}}\right)$$

$$\qquad + \left[\left(\frac{2}{3} \cdot 4^{\frac{3}{2}} - \frac{1}{2} \cdot 4^2 + 2 \cdot 4\right) - \left(\frac{2}{3} \cdot 1^{\frac{3}{2}} - \frac{1}{2} \cdot 1^2 + 2 \cdot 1\right)\right]$$

$$= \left(\frac{4}{3} - 0\right) + \left[\left(\frac{2}{3} \cdot 8 - \frac{1}{2} \cdot 16 + 8\right) - \left(\frac{2}{3} - \frac{1}{2} + 2\right)\right]$$

$$= \frac{4}{3} + \left(\frac{16}{3} - 8 + 8\right) - \left(\frac{4}{6} - \frac{3}{6} + \frac{12}{6}\right)$$

$$= \frac{4}{3} + \frac{16}{3} - \frac{13}{6}$$

$$= \frac{20}{3} - \frac{13}{6}$$

$$= \frac{40}{6} - \frac{13}{6}$$

$$= \frac{27}{6}$$

$$= 4\frac{1}{2}$$

There is an alternative to breaking the integral up into two pieces. The alternative is to estimate the area by breaking the region up into thin *horizontal* strips instead of into thin vertical strips (see Figure 11.25).

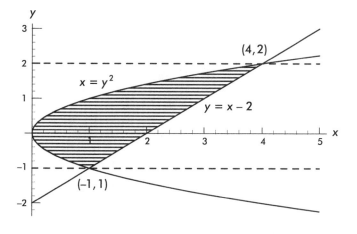

Figure 11.25

Compare Figure 11.25 with Figure 11.4 of Section 11.1. The variable of integration should clearly be y. Each thin horizontal strip is almost a rectangle, with height given by Δy. The length of the strip is the length of a horizontal segment, and is thus given by (x on right) $-$ (x on left). Now the right-hand endpoint of each strip lies on the straight line, and is thus given by $x = y + 2$. The left-hand endpoint of each horizontal strip lies on the curve, and is thus given by $x = y^2$. Thus,

$$(x \text{ on right}) - (x \text{ on left}) = (y + 2) - y^2$$

What are the limits of integration? Remember, y is the variable, so you have to think of two lines of constant y, that is, of two horizontal lines, that contain precisely the region between them. That is, you have to think of the horizontal line that marks off the bottom of the region, and the horizontal line that marks off the top of the region. Clearly, $y = -1$ marks off the bottom, and $y = 2$ marks off the top. Thus, if you think of taking the interval $[-1, 2]$ on the y-axis and breaking it up into n little pieces of length Δy each, and picking a sample point p_i in the ith little piece, and computing

$$\sum_{i=1}^{n} \left((p_i + 2) - (p_i)^2 \right) \Delta y$$

then you would have an estimate of the area of the region. Letting $n \to \infty$, you would get an exact answer. Remember, that in the limit the summation sign becomes an integral, the Δy becomes dy, and the ith sample point p_i just becomes the generic variable, in this case y. Thus the exact area is given by the integral

$$\int_{-1}^{2} [(y+2) - y^2] \, dy$$

$$= \left. \left(\frac{1}{2}y^2 + 2y - \frac{1}{3}y^3 \right) \right|_{-1}^{2}$$

$$= \left(\frac{1}{2} \cdot 4 + 4 - \frac{1}{3} \cdot 8 \right) - \left(\frac{1}{2} \cdot 1 - 2 - \frac{1}{3} \cdot (-1) \right)$$

$$= \left(2 + 4 - \frac{8}{3} \right) - \left(\frac{1}{2} - 2 + \frac{1}{3} \right)$$

$$= \left(6 - 2\frac{2}{3} \right) - \left(\frac{3}{6} - \frac{12}{6} + \frac{2}{6} \right)$$

$$= \left(3\frac{1}{3} \right) - \left(\frac{-7}{6} \right)$$

$$= 3\frac{2}{6} + \frac{7}{6} = 4\frac{1}{2}$$

This is of course the same answer found in the previous example using x as the variable.

There are generally three reasons why you might either have to or prefer to use y as the variable of integration instead of x. One you saw above. Using y might mean that you can get by with only one integral, rather than two. Another reason might be that the curves bounding the region are given in such a way that solving them for y in terms of x requires algebra that is just too hard to do, but solving them for x in terms of y is manageable. A third possibility is that if you use x as the variable, it may be too hard for you to find antiderivatives. Be aware that if this happens, maybe you are supposed to rework the problem and use y as the variable.

As a further example in using y as the variable, let's take one of the earlier examples that you did with x, and redo it with y. The work will look different, but since you are computing the area of the same region, you should of course get the same answer.

Example 11.7 Find the area of the region bounded by $y = (x-1)^2$ and $y = (x-1)^3$, as given in Example 11.3, but use y as the variable.

Solution: In the work in Example 11.3, you already figured out that the curves meet at the points $(1,0)$ and $(2,1)$. Using horizontal instead of vertical strips, so that y is the variable of integration, the picture is that of Figure 11.26.

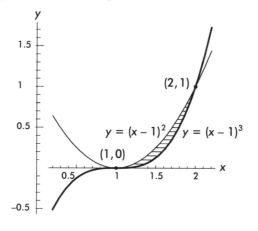

Figure 11.26

Now the height of each horizontal strip is Δy, which becomes dy in the limit. The region is bounded by the x-axis, which is the horizontal line $y = 0$, on the bottom, and on the top by the horizontal line $y = 1$. At each y between 0 and 1, we have a horizontal segment, whose length is the length of the approximating rectangle. The length of the horizontal segment is given by (x on right) − (x on left). As y is the variable of integration, you must solve for x in terms of y. Now the curve on the right is $y = (x-1)^3$, so solving for x you get $x - 1 = y^{1/3}$ or $x = y^{1/3} + 1$, while the curve on the left is $y = (x-1)^2$, so that $x - 1 = y^{1/2}$ or $x = y^{1/2} + 1$. Thus,

$$(x \text{ on right}) - (x \text{ on left}) = \left(y^{\frac{1}{3}} + 1\right) - \left(y^{\frac{1}{2}} + 1\right) = y^{\frac{1}{3}} - y^{\frac{1}{2}}$$

The area of the region is therefore given by

$$
\begin{aligned}
\int_0^1 \left(y^{\frac{1}{3}} - y^{\frac{1}{2}}\right) dy &= \left.\left(\frac{3}{4}y^{\frac{4}{3}} - \frac{2}{3}y^{\frac{3}{2}}\right)\right|_0^1 \\
&= \left(\frac{3}{4} - \frac{2}{3}\right) - (0 - 0) \\
&= \frac{9}{12} - \frac{8}{12} = \frac{1}{12}
\end{aligned}
$$

This is the same answer obtained earlier.

11.3
VOLUMES OF SOLIDS OF REVOLUTION—
THE METHOD OF CROSS-SECTIONAL AREAS

A solid of revolution is the three-dimensional volume you get by rotating a region in the plane about a straight line. For example, when the region in Figure 11.27 is revolved about the y-axis, it sweeps out a solid bowl-shaped volume.

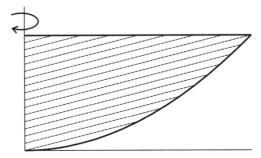

Figure 11.27

If you are having trouble visualizing this, imagine that the region is made out of a thin piece of metal, that the y-axis is a wire, and that the side of the region on the y-axis is hinged to the wire, so that it can revolve around the wire. Now just imagine spinning the region about the wire.

The fundamental principle for computing volume is that if you have a volume in three-space and all the cross sections perpendicular to a line have the same area A, as in either picture of Figure 11.28, then the volume is just $A \cdot h$, the constant cross-sectional area times the length or height of the object.

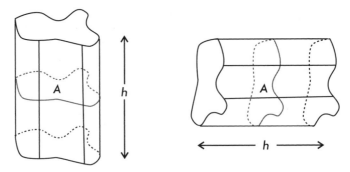

Figure 11.28

This is the principle behind several common volume formulas that you already know, and was discussed in Section 7.1. For example, a right circular cylinder, of base radius r and height h, has cross sections perpendicular to its height that are circles of constant radius r (see Figure 11.29), and thus the cross sections have constant area $A = \pi r^2$. Thus the volume of the cylinder is the constant cross-sectional area πr^2 times the height h, so $V = \pi r^2 h$.

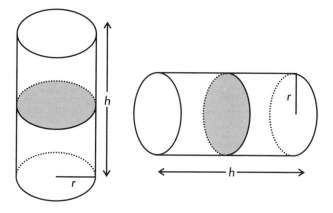

Figure 11.29

What if the cross sections do not have constant area? Well, this is precisely the kind of situation where the idea of the integral can be applied. Recall from Section 8.1 that if you can solve a problem exactly and easily by a multiplication when all the quantities involved are constant, then when one of the quantities is varying, the *idea of the integral* can be applied. Proceeding similarly to the ideas of Section 11.2, you do the following:

1. Break the volume up into lots of little pieces, on each of which the cross-sectional area is almost constant.
2. Remember from Section 11.1 that if your picture shows things broken up by thin vertical strips or by vertical line segments, then the integral must be set up in terms of x, while if your picture shows things broken up by thin horizontal strips or by horizontal line segments, the integral must be set up in terms of y.
3. Estimate the volume of a little piece by assuming that all of the cross-sectional areas in that little piece are constant.
4. Add up the estimates for the volumes of all the little pieces, to get a Riemann or approximate sum estimate for the whole volume.
5. Do this over and over, with thinner and thinner strips, to get the exact answer as a limit of Riemann or approximate sums.
6. The limit of the Riemann or approximate sums is an integral, which by the Fundamental Theorem of Calculus can be computed by finding an antiderivative.

Specifically, suppose again the cross sections do not have constant area, but that you know a formula $A(x)$ for the area of every cross section, at position x, perpendicular to the x-axis. Suppose also that $A(x)$ varies continuously with x (see Figure 11.30).

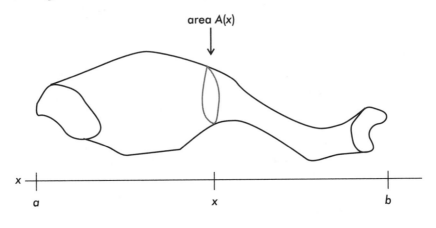

area A(x)

x

a x b

Figure 11.30

The pictured volume lies between $x = a$ and $x = b$. If we break the interval $[a, b]$ on the x-axis up into n small subintervals, the volume V is chopped up into n small pieces also. Use the symbols ΔV_i to denote the volume of the ith little piece, and Δx_i to denote the length of the ith subinterval on the x-axis. If you focus on one particular small subinterval, say the ith, and on the corresponding small chunk of the volume, then the cross-sectional areas in this small chunk of volume do not change very much. Thus you could make the *approximation* or *estimate* that the cross-sectional areas in this small chunk of volume are constant. If you pick a representative point p_i in this ith interval, you can use $A(p_i)$ as a representative value of the cross-sectional area throughout the small chunk of volume, and get

$$\Delta V_i \cong A(p_i) \cdot \Delta x_i$$

as a good estimate for the volume of the ith small chunk (see Figure 11.31). Then

$$V \cong \sum_{i=1}^{n} A(p_i) \cdot \Delta x_i$$

is a good estimate for the whole volume.

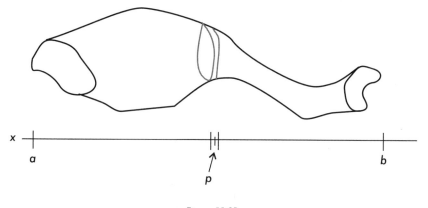

Figure 11.31

The estimates get better and better as you take more and more thinner and thinner pieces, so you get the exact answer in the limit, namely,

$$V = \lim_{n \to \infty} \sum_{i=1}^{n} A(p_i) \cdot \Delta x_i = \int_a^b A(x)\, dx$$

This is true by Definition 8.2 of Section 8.3.
    In other words,

> You can compute volumes by integrating cross-sectional areas.

Note that since it is the x-axis that is being divided into small subintervals, x should be the variable of integration.
    Often, the formula you will be able to find for cross-sectional areas will be for the cross-sectional areas $A(y)$ perpendicular to the y-axis, as in Figure 11.32.

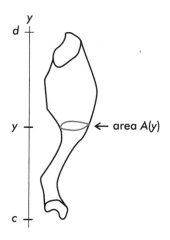

Figure 11.32

In this case, the technique is to break up the y-axis into lots of small subintervals, with the volume correspondingly broken up into lots of small chunks, one of which is pictured in Figure 11.33.

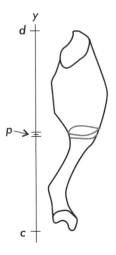

Figure 11.33

If p_i is a representative point in the ith small subinterval on the y-axis, and Δy_i is the length of the ith small subinterval on the y-axis, then

$$\Delta V_i \cong A(p_i) \cdot \Delta y_i$$

So, as before,

$$V = \lim_{n \to \infty} \sum_{i=1}^{n} A(p_i) \cdot \Delta y_i = \int_c^d A(y) \, dy$$

Again, you get volumes by integrating cross-sectional areas, but this time, since it is the y-axis that is being divided into small subintervals, y is the variable of integration, as discussed in Section 11.1.

For solids of revolution, what are the cross-sectional areas? Let's look at some typical cases.

Suppose that $y = f(x)$ is a positive function. Consider the region in the plane bounded above by $y = f(x)$, below by the x-axis, on the left by the vertical line $x = a$, and on the right by the vertical line $x = b$ (see Figure 11.34). If this region is revolved around the x-axis, then, as the cross-sectional areas are swept out by lines *perpendicular to the axis of revolution*, the cross-sectional areas are swept out by vertical line segments.

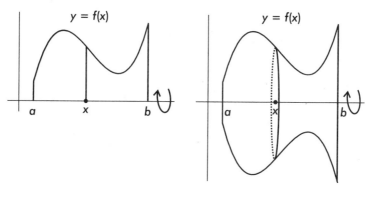

Figure 11.34

The vertical line segment in position x in the figure revolves about the x-axis to sweep out a circle. The area of a circle is πr^2, and the radius r of the cross-sectional circle is precisely the length of the vertical segment at x. By Section 11.1, since you are dealing with vertical line segments, the situation is similar to that depicted in Figure 11.11 of Section 11.1, or in Figures 11.30 and 11.31 of this section. The variable of integration must be x. In terms of x, the vertical line segment has length

$$(y \text{ on top }) - (y \text{ on bottom}) = f(x) - 0 = f(x)$$

Thus the radius of a cross-sectional circle is $r = f(x)$, the area of a cross-sectional circle is $A(x) = \pi(f(x))^2$, and the volume of the solid of revolution is

$$\int_a^b \pi(f(x))^2 \, dx$$

The following situation is slightly more complicated. Suppose that $f(x)$ and $g(x)$ are two positive functions, and that $f(x) \geq g(x)$, so that the graph of f is above that of g. Consider the region in the plane bounded above by $y = f(x)$, below by $y = g(x)$, on the left by the vertical line $x = a$, and on the right by the vertical line $x = b$ (see Figure 11.35). If this region is revolved about the x-axis then, as before, a cross-sectional area is swept out by a vertical line, a line perpendicular to the axis of revolution.

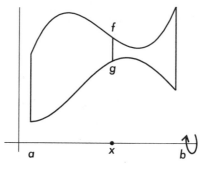

Figure 11.35

In this case, the cross-sectional area looks like a washer, or the region between two concentric circles (that is, between two circles with the same center).

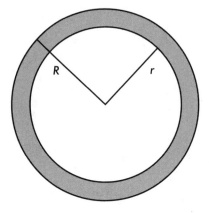

Figure 11.36

If R denotes the radius of the big circle, and r denotes the radius of the small circle, then the area of the washer is the difference between the area of the big circle and the area of the small circle, so the area of the washer is

$$\pi R^2 - \pi r^2$$

That is, the area is (area of the big circle) − (area of the small circle).

The most common mistake for students to make here is to compute the cross-sectional area as $\pi(R - r)^2$. This would give the area of a circle of radius $R - r$, but you can see from the picture above that the cross-sectional area is not a circle.

The radius R of the big circle is the distance between the axis of revolution and the point on the line segment farthest away from the axis, while the radius r of the little circle (the radius of the hole) is the distance between the axis of revolution and the point on the line segment closest to this axis.

As you are breaking the region up by vertical segments, x must be the variable of integration. It follows that $R = f(x)$ and $r = g(x)$, so that the cross-sectional area is $A(x) = \pi(f(x))^2 - \pi(g(x))^2$, and the volume of the solid of revolution is given by

$$V = \int_a^b \pi(f(x))^2 - \pi(g(x))^2 \, dx$$

Example 11.8 The region in the plane bounded above and on the left by the curve $y = x^2$, below by the x-axis, and on the right by the vertical line $x = 5$, is revolved about the x-axis. Find the volume of the resulting solid of revolution.

Solution: The picture is given in Figure 11.37, and includes the drawing of a typical vertical line segment, perpendicular to the axis of rotation, which sweeps out a cross-sectional circle.

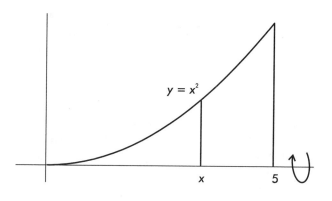

Figure 11.37

As you are breaking the picture up by vertical line segments, the variable of integration must be x. The length of the line segment, which equals the radius of the cross-sectional circle, is

$$(y \text{ on top}) - (y \text{ on bottom}) = x^2 - 0 = x^2$$

So, the cross-sectional area is

$$A(x) = \pi r^2 = \pi (x^2)^2 = \pi x^4$$

The volume is thus

$$
\begin{aligned}
\int_0^5 A(x)\,dx &= \int_0^5 \pi x^4\,dx \\
&= \left. \pi \cdot \frac{1}{5} x^5 \right|_0^5 \\
&= \pi \cdot \frac{1}{5} (5^5 - 0^5) \\
&= \pi \cdot 5^4 = 625\pi
\end{aligned}
$$

Example 11.9 The region in the plane bounded by the curves $y = 2x$ and $y = x^2$ is revolved about the x-axis. Find the volume of the resulting solid of revolution.

Solution: As in Section 11.2, when the vertical lines forming the left- and/or right-hand endpoints of the region are not specified, that means that the curves intersect. The region in question is meant to be the region between the points of intersection, and part of your problem is to figure out what these points are. Setting the formulas equal, you get $2x = x^2$, or

$$x^2 - 2x = x(x - 2) = 0$$

So, the curves intersect at $x = 0$ and at $x = 2$. Substituting these values into the equations, you can see that the points of intersection in the plane are $(0, 0)$ and $(2, 4)$. Picking a point $x = 1$ between 0 and 2, you can see that since $2 \cdot 1 > 1^2$, the graph of $y = 2x$ is above that of $y = x^2$ between $x = 0$ and $x = 2$.

The region, with a sample vertical line segment sweeping out a cross-sectional area, is pictured in Figure 11.38.

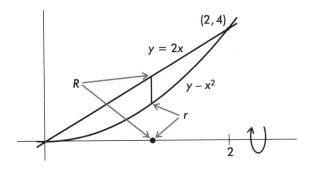

Figure 11.38

As before, with vertical line segments dividing the region, the variable of integration must be x. The cross-sectional area is a washer. The outside radius R of the washer is given by measuring the vertical distance between the point on the vertical line segment *farthest* from the axis of revolution and the axis of revolution. Clearly $R = 2x - 0 = 2x$. The radius r of the smaller circle, that is, the radius r of the hole in the washer, is given by measuring the vertical distance between the point on the vertical line segment *closest* to the axis of revolution and the axis of revolution. This is $r = x^2 - 0 = x^2$. Thus the cross-sectional area is

$$A(x) = \pi R^2 - \pi r^2 = \pi(2x)^2 - \pi(x^2)^2 = \pi(4x^2 - x^4)$$

The volume of the solid of revolution is then given by

$$
\begin{aligned}
\int_0^2 A(x) &= \int_0^2 \pi(4x^2 - x^4)\,dx \\
&= \pi\left(\frac{4}{3}x^3 - \frac{1}{5}x^5\right)\Big|_0^2 \\
&= \pi\left[\left(\frac{4}{3}\cdot 2^3 - \frac{1}{5}\cdot 2^5\right) - \left(\frac{4}{3}\cdot 0^3 - \frac{1}{5}\cdot 0^5\right)\right] \\
&= \pi\left(\frac{32}{3} - \frac{32}{5}\right) \\
&= \pi\left(\frac{32\cdot 5 - 32\cdot 3}{15}\right) = \frac{64}{15}\pi
\end{aligned}
$$

If a region in the plane is revolved about the y-axis, or about any vertical line, the approach is similar. The main difference is that since the line segment that sweeps out a cross-sectional area must be perpendicular to the axis of revolution, when the axis of revolution is vertical, the typical line segment must be horizontal. As you can see from Figure 11.4 of Section 11.1, and from Figures 11.32 and 11.33 of this section, the variable of integration must be y. To get everything set up in terms of y, either the boundaries of the region will be given to you as functions where x is expressed in terms of y, that is, in the form $x = g(y)$, or they will be given to you in the usual form $y = f(x)$, and part of your job in solving the problem is to rewrite the equation to get x in terms of y.

Example 11.10 The region in the plane bounded by the curves $y = 2x$ and $y = x^2$ is revolved about the y-axis. Find the volume of the resulting solid of revolution.

Solution: Note that this is the same region in the plane as in the previous example, but it is being revolved about a different axis. Thus you are getting a different solid of revolution, and there is no reason whatsoever to expect the same answer as in the previous example. Now a sample line segment sweeping out a cross-sectional area is horizontal, as pictured in Figure 11.39.

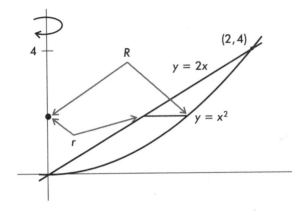

Figure 11.39

The integral must be set up in terms of the variable y. The first thing to notice is that the limits of integration are different from those in the previous example. The region you are revolving has its bottom point on the line $y = 0$ and its top point on the line $y = 4$. Thus the interval on the y-axis that you would divide into small subintervals is the interval $[0, 4]$, and because of that, the limits of integration, with y as the variable, should be from 0 to 4.

The horizontal line segment sweeps out a washer as it revolves about the y-axis. The outside radius R of the washer is the distance between the point on the horizontal line segment *farthest* away from the axis of revolution to the axis of revolution. It is thus the distance between the point on the curve $y = x^2$ to the y-axis. The length of a horizontal line is given by

$$(x \text{ on the right }) - (x \text{ on the left})$$

However, as mentioned above, this must be computed in terms of y. Now if $y = x^2$, then $x = \sqrt{y}$. So the x on the right is $\sqrt{y}$. The x on the left is a point on the y-axis, which is the line $x = 0$. Thus,

$$R = \sqrt{y} - 0 = \sqrt{y}$$

The radius of the inner circle, that is, the radius of the hole, is given by measuring the distance between the point on the horizontal line segment *closest* to the axis of revolution and the axis of revolution. It is thus the distance between the point on the curve $y = 2x$ and the y-axis. You measure this horizontal distance by

$$(x \text{ on the right }) - (x \text{ on the left})$$

As before, x must be computed in terms of y. As $y = 2x$, it is clear that $x = \dfrac{1}{2}y$. Thus,

$$r = \frac{1}{2}y - 0 = \frac{1}{2}y$$

It follows that the cross-sectional area is

$$A(y) = \pi(R^2 - r^2) = \pi\left((\sqrt{y})^2 - \left(\frac{1}{2}y\right)^2\right) = \pi\left(y - \frac{1}{4}y^2\right)$$

The volume of the solid is given by

$$
\begin{aligned}
\int_0^4 A(y)\,dy &= \int_0^4 \pi\left(y - \frac{1}{4}y^2\right) dy \\[2mm]
&= \left.\pi\left(\frac{1}{2}y^2 - \frac{1}{12}y^3\right)\right|_0^4 \\[2mm]
&= \pi\left[\left(\frac{1}{2}\cdot 4^2 - \frac{1}{12}\cdot 4^3\right) - \left(\frac{1}{2}\cdot 0^2 - \frac{1}{12}\cdot 0^3\right)\right] \\[2mm]
&= \pi\left[\left(8 - \frac{64}{12}\right) - 0\right] \\[2mm]
&= \pi\left(8 - \frac{16}{3}\right) = \frac{8\pi}{3}
\end{aligned}
$$

Similar methods will allow you to find the volumes of solids of revolution formed by revolving a region in the plane not only about the x-axis or the y-axis, but about *any* horizontal or vertical line.

Example 11.11   The region in the plane bounded below and on the right by the curve $y = x^2$, on the left by the y-axis, and on top by the line $y = 4$, is revolved about the line $y = 4$. Find the volume of the resulting solid of revolution.

Solution: The axis of revolution is a horizontal line, so that vertical line segments perpendicular to the axis of revolution sweep out cross-sectional areas. With vertical line segments, the variable of integration must be x (see Figure 11.40).

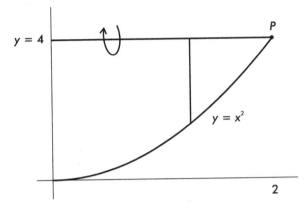

Figure 11.40

Note that to figure out the interval on the x-axis over which the picture lies, you have to compute the x-coordinate of the point P. As P lies on the curve, its x and y coordinates are related by the formula of the curve, $y = x^2$. As P also lies on the line $y = 4$, the y coordinate of P equals 4. Thus to get the x-coordinate of P, you have to substitute 4 for y in the equation $y = x^2$, and solve for x. So solve $4 = x^2$ for x. There are two solutions, namely, $x = 2$ and $x = -2$. As P is clearly to the right of the y-axis, its x-coordinate is positive, and thus $x = 2$, so that $P = (2, 4)$. Thus it is the interval $[0, 2]$ on the x-axis that is being broken up into little pieces, and the limits of integration, with x as the variable of integration, will be from 0 to 2.

As one end of the vertical line segment is on the axis of revolution, the line segment sweeps out a circle, not a washer. The length of the line segment is the radius of the circle, and is given by (y on top) − (y on bottom.) This must be expressed in terms of x. The top of the vertical line segment lies on the line $y = 4$. The bottom lies on the curve $y = x^2$. Thus the radius of a cross-sectional circle is

$$R = (y \text{ on top }) - (y \text{ on bottom}) = 4 - x^2$$

The area of a cross-sectional circle is then

$$A(x) = \pi R^2 = \pi(4 - x^2)^2 = \pi(16 - 8x^2 + x^4)$$

Finally, the volume of the solid of revolution is

$$\int_0^2 A(x)\, dx \quad = \quad \int_0^2 \pi(16 - 8x^2 + x^4)\, dx$$

$$= \pi \left(16x - \frac{8}{3}x^3 + \frac{1}{5}x^5 \right) \Big|_0^2$$

$$= \pi \left[\left(32 - \frac{64}{3} + \frac{32}{5} \right) - (0 - 0 + 0) \right]$$

$$= \pi \left(\frac{32 \cdot 15 - 64 \cdot 5 + 32 \cdot 3}{15} \right)$$

$$= \pi \left(\frac{480 - 320 + 96}{15} \right) = \frac{256\pi}{15}$$

Example 11.12 The region bounded below and on the right by the curve $y = x^2$, on the top by the line $y = 4$, and on the left by the line $x = 1$, is revolved about the line $x = -1$. Find the volume of the resulting solid of revolution.

Solution: I changed the left boundary of the region from the y-axis to the line $x = 1$ just for a little variety. Since you are revolving about a vertical line, horizontal line segments sweep out the cross-sectional areas, and the variable of integration must be y (see Figure 11.41).

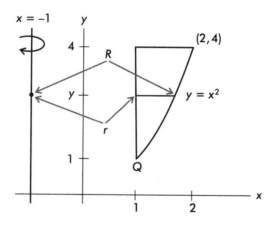

Figure 11.41

The top point of the region being revolved is on the the line $y = 4$, so $y = 4$ is the upper limit of integration. The bottom of the region is at the point Q. As the x-coordinate of Q equals 1 and Q lies on the curve $y = x^2$, then its y-coordinate is $1^2 = 1$, and the bottom limit of integration is $y = 1$.

As the horizontal segment in the region being rotated does *not* have one of its endpoints on the axis of revolution, the cross-sectional area being swept out is a washer. The length of R is the horizontal distance between the point on the segment farthest away from the axis of revolution and the axis of revolution. It is of the form (x on right) − (x on left), but must be computed in terms of y. The point on the segment farthest from the axis of revolution is on the curve $y = x^2$, so that $x = \sqrt{y}$. The axis of revolution is the line $x = -1$. Thus,

$$R = (x \text{ on right}) - (x \text{ on left}) = \sqrt{y} - (-1) = \sqrt{y} + 1$$

The radius r of the inside circle, or hole in the washer, is the distance between the point on the segment closest to the axis of revolution and the axis of revolution. The point on the segment closest to the axis of revolution lies on the line $x = 1$. Thus,

$$r = (x \text{ on right}) - (x \text{ on left}) = 1 - (-1) = 2$$

The cross-sectional area is

$$
\begin{aligned}
A(y) = \pi(R^2 - r^2) &= \pi[(\sqrt{y} + 1)^2 - 2^2] \\
&= \pi[(y + 2\sqrt{y} + 1) - 4] \\
&= \pi\left(y + 2y^{\frac{1}{2}} - 3\right)
\end{aligned}
$$

The volume of the solid of revolution is

$$
\begin{aligned}
\int_1^4 A(y)\, dy &= \int_1^4 \pi\left(y + 2y^{\frac{1}{2}} - 3\right) dy \\
&= \pi\left(\frac{1}{2}y^2 + 2 \cdot \frac{2}{3}y^{\frac{3}{2}} - 3y\right)\Big|_1^4 \\
&= \pi\left[\left(\frac{1}{2} \cdot 16 + \frac{4}{3}(4^{\frac{1}{2}})^3 - 3 \cdot 4\right)\right. \\
&\quad \left. - \left(\frac{1}{2} \cdot 1 + \frac{4}{3} \cdot 1 - 3 \cdot 1\right)\right] \\
&= \pi\left[\left(8 + \frac{4}{3} \cdot 2^3 - 12\right) - \left(\frac{1}{2} + \frac{4}{3} - 3\right)\right] \\
&= \pi\left[\left(\frac{32}{3} - 4\right) - \left(\frac{3}{6} + \frac{8}{6} - 3\right)\right] \\
&= \pi\left(\frac{32}{3} - 4 - \frac{11}{6} + 3\right) \\
&= \pi\left(\frac{64}{6} - \frac{11}{6} - 1\right) = \frac{47\pi}{6}
\end{aligned}
$$

Review of Main Points of This Section

To compute volumes of solids of revolution, by the method of cross-sectional areas:

- Sketch the curves bounding the region that is being revolved.

- Draw a curving arrow around the axis of revolution, so that as you work the problem, you won't forget which line it is.

- Draw a horizontal or vertical segment in the region being revolved—for the method of cross-sectional areas, your segments should be perpendicular to the axis of revolution.

- Remember that if your line segments are horizontal then y is the variable of integration, and if they are vertical, x is the variable of integration.

- If one end of the line segment lies on the axis of revolution, the cross sections are circles; otherwise the cross sections are washers.

- If the cross sections are circles, write down a formula for r and integrate πr^2.

- If the cross sections are washers, write down a formula for R and r and integrate $\pi R^2 - \pi r^2$.

- Remember from the first section of this chapter how to measure the lengths of line segments.

- Be sure to measure the lengths of line segments in terms of the correct variable of integration.

- If your variable of integration is x, the limits of integration are the x-coordinates of the leftmost and rightmost points in the region being revolved; if your variable of integration is y, the limits of integration are the y-coordinates of the bottom and top points in the region being revolved; the axis of revolution does not affect the limits of integration.

11.4 VOLUMES OF SOLIDS OF REVOLUTION —THE METHOD OF CYLINDRICAL SHELLS

Suppose you were given the problem of finding the volume of the solid of revolution formed when the region in the plane bounded above by the curve $y = -x^2 + 3x - 2 = -(x - 1)(x - 2)$ and below by the x-axis, is revolved about the y-axis. A horizontal line segment in the region, perpendicular to the axis of revolution, sweeps out a cross-sectional washer (see Figure 11.42). To compute R and r in terms of the variable of integration y, you would have to take the quadratic $y = -x^2 + 3x - 2$ and find the two solutions for x in terms of y.

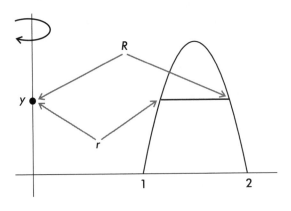

Figure 11.42

While doing the algebra is possible, it seems like an awful nuisance. It certainly would be nice if there were an easier way to do the problem, using the given equation $y = -x^2 + 3x - 2$ directly. There is, and it is called the **method of cylindrical shells**.

If you draw, inside the region being revolved, a *thin vertical* strip, *parallel* to the axis of revolution, it sweeps out what is called a **cylindrical shell**, that is, the curved surface of a cylinder, but with a little thickness to it, so that it is a three-dimensional volume, and not just a two-dimensional area (see Figure 11.43).

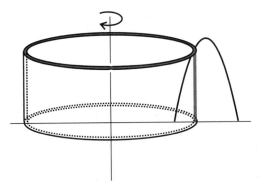

Figure 11.43

Of course, since the vertical strip is not *exactly* a rectangle, the volume it sweeps out is not *exactly* a cylindrical shell. However, the thin strip is *approximately* a rectangle, and the volume it sweeps out is *approximately* a cylindrical shell. Furthermore, as the vertical strip gets thinner, it is closer to being a rectangle, and the volume it sweeps out is closer to being a cylindrical shell. Make the estimate or approximation that the thin strip is a rectangle, and that the volume is a cylindrical shell.

If you divided the whole region into thin vertical strips, then the volume of the solid of revolution would be the sum of the volumes of the *approximate* cylindrical shells being swept out by the strips. If you knew the formula for the volume ΔV_i of a cylindrical shell, you would have an approximate sum $V = \Sigma_{i=1}^{n} \Delta V_i$, with the approximations getting better and better as the intervals get thinner and thinner.

Let's see how to get the formula for the volume of a cylindrical shell. Suppose that a thin vertical strip is being revolved about a vertical line to form a cylindrical shell. The height h of the strip is the height h of the shell, and the distance from the strip to the axis of revolution is the radius r of the shell. The idea of how to estimate the volume of the shell is similar to what we did in Section 7.1, Figure 7.2, to get the area of the curved surface of a cylinder. Imagine cutting the cylindrical shell along a line parallel to the axis of revolution, and spreading it out.

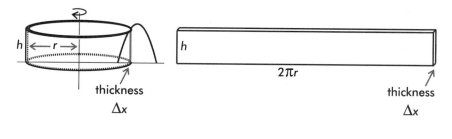

Figure 11.44

You will more or less get a rectangle, *but with a little thickness*—in other words, a three-dimensional box (approximately). The volume of the box is its length times its height times its width. Now the height of the box is just the height h of the cylinder, while the length of the box is the circumference $2\pi r$ of the cylinder. The *width* of the box is the width of the cylindrical shell, which is nothing other than the width of the original vertical strip being revolved. You shouldn't be surprised if we call this width Δx, since it represents the width of a small subinterval of the x-axis. Thus,

$$\Delta V = h \cdot 2\pi r \cdot \Delta x$$

and

$$V \text{ is approximately equal to } \sum_{i=1}^{n} h \cdot 2\pi r \cdot \Delta x$$

Of course, the estimates get better and better as the intervals get thinner and thinner, so you get the *exact* volume in the limit,

$$V = \lim_{n \to \infty} \sum_{i=1}^{n} h \cdot 2\pi r \cdot \Delta x = \int h \cdot 2\pi r \, dx$$

Of course, you must compute h and $2\pi r$ in terms of the variable of integration, which in the above problem is x. But the moral is this:

> You can compute volumes by integrating the areas of curved cylindrical surfaces.

If, in the limit, you imagine that the thin vertical strip becomes a thin vertical line, then the line sweeps out a curved cylindrical surface, and it is precisely the area of this surface that you integrate. You should be able to visualize the line sweeping out the surface, and to visualize and label the radius r and height h of the surface, as shown in Figure 11.45.

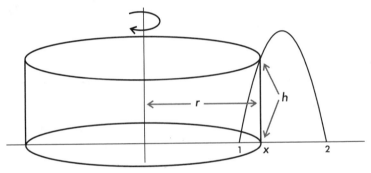

Figure 11.45

The same idea applies for revolving about a horizontal line as shown in Figure 11.46.

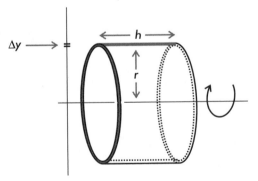

Figure 11.46

The thin horizontal strip sweeps out a cylindrical shell. The length of the strip is the height h of the cylinder, the distance between the strip and the axis of revolution is the radius r of the cylinder, and the thickness Δy of the strip is the thickness of the cylindrical shell. It follows as before that the volume of the cylindrical shell is

$$\Delta V = h \cdot 2\pi r \Delta y$$

Exactly as before with revolving a region about the y-axis, if you were trying to compute the volume of the solid of revolution formed by rotating a region about the x-axis, by the method of cylindrical shells, you would draw inside the region a *horizontal* line *parallel* to the axis of revolution. Imagine the horizontal line sweeping out the curved surface of a cylinder. The curved surface area $2\pi rh$, computed in terms of the variable of integration y, is what you would integrate. Identify in your picture what r and h are. Thus, if you were trying to compute the volume of the solid of revolution formed by rotating the region pictured in Figure 11.47 about the x-axis, the answer would be

$$V = \int_c^d 2\pi rh \, dy$$

where again you would have to compute r and h in terms of the variable of integration y.

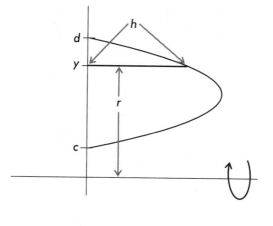

Figure 11.47

Example 11.13 The region in the plane bounded below by the x-axis, and bounded above by the curve $y = -x^2 + 3x - 2$ is revolved about the y-axis. Find the volume of the resulting solid of revolution.

Solution: You first have to do a little algebra to find the limits of integration. Factor the quadratic

$$-x^2 + 3x - 2 = -(x^2 - 3x + 2) = -(x - 2)(x - 1)$$

to see that the quadratic crosses the x-axis at $x = 1$ and at $x = 2$, and is above the x-axis between $x = 1$ and $x = 2$. A vertical line, *parallel* to the axis of revolution, sweeps out a cylinder (see Figure 11.48). The vertical line means that x must be the variable of integration. The surface area of the cylinder, which is what you integrate, is $2\pi r h$. Now you can see that the height of the cylinder is the length of the vertical line, that is,

$$(y \text{ on top }) - (y \text{ on bottom}) = (-x^2 + 3x - 2) - 0 = -x^2 + 3x - 2$$

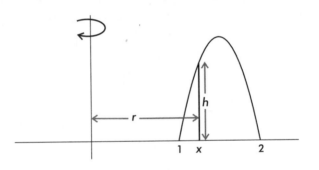

Figure 11.48

What about the radius of the cylinder? Well, that's just the distance between the general vertical line at position x and the y-axis $x = 0$. Thus the radius is the length of the horizontal line between x and 0, namely $x - 0 = x$. It follows that

$$2\pi r h = 2\pi x \cdot (-x^2 + 3x - 2) = 2\pi(-x^3 + 3x^2 - 2x)$$

Thus,

$$
\begin{aligned}
V = \int_1^2 2\pi(-x^3 + 3x^2 - 2x)\, dx \;&=\; 2\pi\left(-\frac{1}{4}x^4 + x^3 - x^2\right)\Big|_1^2 \\
&=\; 2\pi\left[\left(-\frac{1}{4}\cdot 16 + 8 - 4\right)\right. \\
&\qquad\left. -\left(-\frac{1}{4}\cdot 1 + 1 - 1\right)\right] \\
&=\; 2\pi\left[(-4 + 8 - 4) - \left(\frac{-1}{4}\right)\right] = \frac{\pi}{2}
\end{aligned}
$$

For the rest of this section, we'll do the examples of the previous section, but by the method of cylindrical shells.

Example 11.14 The region in the plane bounded above and on the left by the curve $y = x^2$, below by the x-axis, and on the right by the vertical line $x = 5$, is revolved about the x-axis. Find, by the method of cylindrical shells, the volume of the resulting solid of revolution.

Solution: Draw in a line segment *parallel* to the axis of revolution as in Figure 11.49. For this problem, that means a horizontal line segment, which implies that the variable of integration must be y.

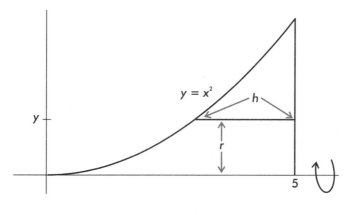

Figure 11.49

The bottom of the region being revolved lies on the x-axis, that is, on the line $y = 0$, while the top point of the region has y-coordinate $y = x^2 = 5^2 = 25$. Thus the limits of integration are from 0 to 25. The height h of the cylinder is the length of the horizontal line segment, which is given by

$$(x \text{ on right}) - (x \text{ on left}) = 5 - \sqrt{y} = 5 - y^{\frac{1}{2}}$$

The radius of the cylinder is the distance between the general horizontal line at position y and the axis of revolution $y = 0$, that is, $y - 0 = y$. Thus the area of the curved surface of the cylinder swept out by the horizontal line segment at position y equals

$$2\pi rh = 2\pi y(5 - y^{\frac{1}{2}}) = 2\pi(5y - y^{\frac{3}{2}})$$

The volume is given by

$$
\begin{aligned}
V = \int_0^{25} 2\pi(5y - y^{\frac{3}{2}})\, dy &= 2\pi\left(\frac{5}{2}y^2 - \frac{2}{5}y^{\frac{5}{2}}\right)\Bigg|_0^{25} \\
&= 2\pi\left[\left(\frac{5}{2}\cdot 625 - \frac{2}{5}(25^{\frac{1}{2}})^5\right) - (0 - 0)\right] \\
&= 2\pi\left(\frac{3{,}125}{2} - \frac{2}{5}\cdot 5^5\right) \\
&= 2\pi\left(\frac{3{,}125}{2} - 2\cdot 5^4\right) \\
&= \pi(3{,}125 - 4\cdot 625) = \pi(3{,}125 - 2{,}500) = 625\pi
\end{aligned}
$$

This is the same answer obtained in Example 11.8 of the previous section. It should of course be the same answer, since you are computing, by a different method, the volume of the same solid.

Example 11.15 The region in the plane bounded by the curves $y = 2x$ and $y = x^2$ is revolved about the x-axis. Find, by the method of cylindrical shells, the volume of the resulting solid of revolution.

Solution: A *horizontal* line segment sweeps out a cylindrical surface. You already saw in Example 11.9 of the previous section that the curves meet at $(0, 0)$ and at $(2, 4)$. Thus the y limits of integration are from 0 to 4. The height of the cylinder is the length of the horizontal segment, computed in terms of y (see Figure 11.50). The right-hand endpoint of the segment lies on $y = x^2$, so that (x on right) $= \sqrt{y} = y^{\frac{1}{2}}$. The left-hand endpoint of the segment lies on $y = 2x$, so that (x on left) $= \dfrac{y}{2}$. Thus the height h of the cylinder is

$$
h = (x \text{ on right}) - (x \text{ on left}) = y^{\frac{1}{2}} - \frac{y}{2}
$$

The radius of the cylinder is y, exactly as in the previous example.

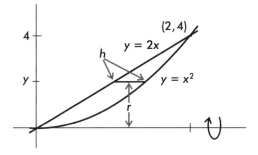

Figure 11.50

Thus the surface area of the cylinder is

$$2\pi rh = 2\pi y\left(y^{\frac{1}{2}} - \frac{y}{2}\right) = 2\pi\left(y^{\frac{3}{2}} - \frac{1}{2}y^2\right)$$

The volume equals

$$
\begin{aligned}
\int_0^4 2\pi\left(y^{\frac{3}{2}} - \frac{1}{2}y^2\right) dy &= 2\pi\left(\frac{2}{5}y^{\frac{5}{2}} - \frac{1}{6}y^3\right)\Big|_0^4 \\
&= 2\pi\left[\left(\frac{2}{5}\cdot(4^{\frac{1}{2}})^5 - \frac{1}{6}\cdot 4^3\right) - (0-0)\right] \\
&= 2\pi\left(\frac{2}{5}\cdot 2^5 - \frac{1}{6}\cdot 64\right) \\
&= 2\pi\left(\frac{64}{5} - \frac{64}{6}\right) \\
&= 2\pi\left(\frac{64\cdot 6 - 64\cdot 5}{30}\right) \\
&= 2\pi\cdot\frac{64}{30} \\
&= \frac{64\pi}{15}
\end{aligned}
$$

Again, this is the same answer obtained in Example 11.9.

Example 11.16 The region in the plane bounded by the curves $y = 2x$ and $y = x^2$ is revolved about the y-axis. Find the volume of the resulting solid of revolution, by the method of cylindrical shells.

Solution: In this example it is a *vertical* line segment that is parallel to the axis of revolution, and that sweeps out a cylindrical surface. Thus the variable of integration is x. As the leftmost point of the region being revolved has coordinates $(0, 0)$ and the rightmost point of the region has coordinates $(2, 4)$, the x limits of integration are from 0 to 2, just as in Example 11.9 of the previous section. The height h of a vertical segment, in terms of x, is just $2x - x^2$, and that is also the height h of the cylindrical surface (see Figure 11.51). The radius r of the cylinder is the distance between the vertical segment, at general position x, and the axis of revolution $x = 0$, so the radius $r = x - 0 = x$.

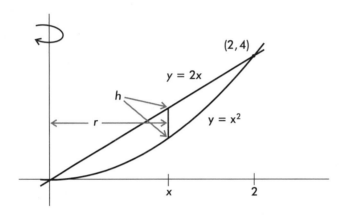

Figure 11.51

The surface area of the cylinder swept out by the vertical line segment is thus

$$2\pi r h = 2\pi x(2x - x^2) = 2\pi(2x^2 - x^3)$$

The volume is therefore

$$
\begin{aligned}
V = \int_0^2 2\pi(2x^2 - x^3)\, dx &= 2\pi\left(\frac{2}{3}x^3 - \frac{1}{4}x^4\right)\Bigg|_0^2 \\
&= 2\pi\left[\left(\frac{2}{3}\cdot 2^3 - \frac{1}{4}\cdot 2^4\right) - (0 - 0)\right] \\
&= 2\pi\left(\frac{16}{3} - 4\right) \\
&= 2\pi\left(\frac{16 - 12}{3}\right) = \frac{8\pi}{3}
\end{aligned}
$$

Again, this is the same answer as found in Example 11.10 of the previous section. In this example, it seemed to be easier to use the method of cylindrical shells than the method of cross-sectional areas.

Example 11.17 The region in the plane bounded below and on the right by the curve $y = x^2$, on the left by the y-axis, and on top by the line $y = 4$, is revolved about the line $y = 4$. Find the volume of the resulting solid of revolution by the method of cylindrical shells.

Solution: A cylindrical shell is swept out by a line segment *parallel* to the axis of revolution. For this problem, that means a horizontal line segment, so the variable of integration should be y. The bottom of the region is at $y = 0$, while the top of the region being revolved is at $y = 4$, so the y limits of integration are from 0 to 4. The height h of the cylinder is the length of the horizontal line segment in the region (see Figure 11.52). This is given by the x-coordinate of the right-hand endpoint minus the x-coordinate of the left-hand endpoint. You must find these x-coordinates in terms of the variable of integration y. As the right-hand endpoint of the horizontal line segment is on the curve $y = x^2$, (x on right) $= \sqrt{y} = y^{\frac{1}{2}}$. The left-hand endpoint of the horizontal segment is on the y-axis, that is, on the line $x = 0$. Thus the height $h = y^{\frac{1}{2}} - 0 = y^{\frac{1}{2}}$. The radius of the cylinder is the distance between the general horizontal segment at position y, and the line around which it is being revolved, which in this example is $y = 4$. Computing this vertical distance by (y on top) $-$ (y on bottom), you get that $r = 4 - y$.

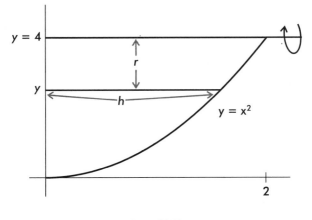

Figure 11.52

The area of the cylindrical surface is thus

$$2\pi rh = 2\pi(4 - y)(y^{\frac{1}{2}}) = 2\pi(4y^{\frac{1}{2}} - y^{\frac{3}{2}})$$

The volume equals

$$
\begin{aligned}
V = \int_0^4 2\pi \left(4y^{\frac{1}{2}} - y^{\frac{3}{2}}\right) dy
&= 2\pi \left(4 \cdot \frac{2}{3} y^{\frac{3}{2}} - \frac{2}{5} y^{\frac{5}{2}}\right)\Big|_0^4 \\
&= 2\pi \left[\left(\frac{8}{3} \cdot 4^{\frac{3}{2}} - \frac{2}{5} \cdot 4^{\frac{5}{2}}\right) - (0 - 0)\right] \\
&= 2\pi \left(\frac{8}{3}(4^{\frac{1}{2}})^3 - \frac{2}{5}(4^{\frac{1}{2}})^5\right) \\
&= 2\pi \left(\frac{8}{3} \cdot 8 - \frac{2}{5} \cdot 32\right) \\
&= 2\pi \left(\frac{64 \cdot 5 - 64 \cdot 3}{15}\right) \\
&= 2\pi \cdot \frac{128}{15} = \frac{256\pi}{15}
\end{aligned}
$$

The work is quite different, but the answer is the same as that for Example 11.11 in the previous section.

Example 11.18 The region bounded below and on the right by the curve $y = x^2$, on the top by the line $y = 4$, and on the left by the line $x = 1$, is revolved about the line $x = -1$. Find the volume of the resulting solid of revolution, by the method of cylindrical shells.

Solution: A vertical line, parallel to the axis of revolution, sweeps out a cylindrical surface. The variable of integration must therefore be x. The leftmost part of the region being revolved is on the line $x = 1$. The rightmost point of the region is the point P, which is on the curve $y = x^2$ and on the line $y = 4$. Substituting 4 for y in $y = x^2$ and solving for x gives $x = 2$. Thus the rightmost point of the region being revolved has x-coordinate 2, and the limits of integration, in terms of x, are from 1 to 2. The height h of the cylinder is the height of the vertical strip, which is (y on top) $-$ (y on bottom) or, in terms of x, $4 - x^2$ (see Figure 11.53). The radius of the cylinder is the horizontal distance between the vertical line being revolved, and the axis of revolution. The vertical line being revolved is to the right of the axis of revolution and has general position x, while the axis of revolution is on the left and has position $x = -1$. Thus the radius is $x - (-1) = x + 1$.

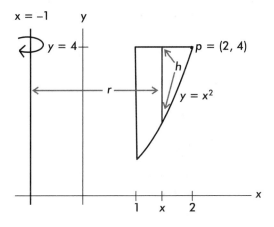

Figure 11.53

The curved surface of the cylinder is thus

$$2\pi r h = 2\pi(x+1)(4-x^2) = 2\pi(4x+4-x^3-x^2)$$

The volume of the solid of revolution equals

$$
\begin{aligned}
V = \int_{1}^{2} 2\pi(4x+4-x^3-x^2)\,dx \;&=\; 2\pi\left(2x^2+4x-\frac{1}{4}x^4-\frac{1}{3}x^3\right)\Big|_{1}^{2}\\[6pt]
&=\; 2\pi\left[\left(8+8-\frac{1}{4}\cdot16-\frac{1}{3}\cdot8\right)\right.\\[6pt]
&\qquad\left.-\left(2+4-\frac{1}{4}-\frac{1}{3}\right)\right]\\[6pt]
&=\; 2\pi\left[\left(12-\frac{8}{3}\right)-\left(6-\frac{1}{4}-\frac{1}{3}\right)\right]\\[6pt]
&=\; 2\pi\left(12-\frac{8}{3}-6+\frac{1}{4}+\frac{1}{3}\right)\\[6pt]
&=\; 2\pi\left(6-\frac{7}{3}+\frac{1}{4}\right)\\[6pt]
&=\; 2\pi\left(\frac{6\cdot12-7\cdot4+1\cdot3}{12}\right)\\[6pt]
&=\; 2\pi\left(\frac{72-28+3}{12}\right)\\[6pt]
&=\; 2\pi\cdot\frac{47}{12}=\frac{47\pi}{6}
\end{aligned}
$$

This is exactly as in Example 11.12 of the previous section.

11.5 SUMMARY OF MAIN POINTS

- Whether you are computing areas or volumes of solids of revolution, you always start out with a region in the plane, described by the curves that form its boundaries. Sketch the curves and identify the region.

  — Sketching the region accurately may require some algebra to figure out where the curves cross. If this is the case, set the formulas equal, and solve. This will give you one of the coordinates of the crossing points. Substitute this coordinate into the formula to get the other coordinate of the crossing points.

  — Between crossing points, evaluate both formulas to make sure you know which curve is on top and which is on the bottom.

- Whether you are doing an area or a volume problem, look at the picture of the region, and look at the formulas for the boundaries, and think about whether it might be easier to use x or y as the variable of integration. Of course, if you are doing a volume problem and the problem specifies which method to use (cross-sectional areas or cylindrical shells) then you do not have an option.

- Remember that x as the variable of integration corresponds to filling up the region with sample *vertical* line segments.

- Remember that y as the variable of integration corresponds to filling up the region with sample *horizontal* line segments.

- Always actually draw inside the region a sample line segment.

- The following are some reasons why one variable of integration might be easier to use than another.

  — The choice of one variable of integration might require solving the problem by setting up two separate integrals. This would happen if, in different parts of the region, sample line segments would have endpoints on different curves. Using the other variable of integration might allow you to solve the problem with only one integral.

  — The choice of one variable of integration might require you to do some difficult algebra. For example, if x is the variable of integration, you will generally have to take the formulas for the boundaries of the region, and solve for y in terms of x. Conversely, if y is the variable of integration, you will generally have to take the formulas for the boundaries of the region, and solve for x in terms of y. One of these choices might be fairly easy, while the other might be difficult.

— It could be that with one variable of integration, you end up with an integrand that is difficult to find the antiderivative of, but with the other variable of integration, the integrand is simple. Often you cannot tell at the beginning of the problem, just by looking at the picture of the region, if this might happen.

- If x is your variable of integration, the lower limit of integration is the x-coordinate of the leftmost point of the region, and the upper limit of integration is the x-coordinate of the rightmost point of the region. If y is your variable of integration, the lower limit of integration is the y-coordinate of the lowest point of the region, and the upper limit of integration is the y-coordinate of the highest point of the region. In particular, for volumes of solids of revolution, the limits of integration are determined solely by the region being revolved, and not at all by the axis of revolution.

- Remember that you compute the length of a vertical line by

$$(y \text{ on top }) - (y \text{ on bottom})$$

You will often have to find these y's in terms of x.

- Remember that you compute the length of a horizontal line by

$$(x \text{ on right }) - (x \text{ on left})$$

You will often have to find these x's in terms of y.

- For an area problem, the integrand is simply the length of a sample line segment.

- For a volume problem, always draw in a curved arrow around the axis of revolution, so that your picture gives you good visual information as to what is going on.

- If your sample line segment is *perpendicular* to the axis of revolution, you are using the method of *cross-sectional areas*. Your integrand is the cross-sectional area swept out by the sample line segment, and this will be a circle or a washer.

  — If throughout the region one end of the line segment is *always* on the axis of revolution, your cross section is a circle. The integrand is πr^2, where r, the radius of the circle, is simply the length of the line segment.

  — If one end of the line segment is not always on the axis of revolution, your cross section is a washer, the region between two circles, and the integrand is $\pi R^2 - \pi r^2$. The radius R of the large circle is the distance between the axis of revolution and the point on the sample line segment *farthest* from the axis of revolution. The radius

r of the small circle, or hole in the washer, is the distance between the axis of revolution and the point on the sample line segment *closest* to the axis of revolution.

- If your sample line segment is *parallel* to the axis of revolution, you are using the method of *cylindrical shells*. Your integrand is the curved surface area $2\pi rh$ of the cylinder swept out by the sample line segment. The height h of the cylinder is the same as the length of the sample line segment, while the radius r of the cylinder is the distance between the sample line segment and the axis of revolution.

11.6 EXERCISES

In Exercises 1–10, find the area of the region in the plane bounded, as indicated, by the given curves. For each problem, sketch the graph of the region, and draw in a sample line segment.

1. The region bounded below by $y = x$, above by $y = x^2$, on the left by the vertical line $x = 2$, and on the right by the vertical line $x = 4$.

2. The region between $y = x^2 - 1$ and the x-axis.

3. The region bounded by the curves $y = x$ and $y = x^2$, between $x = 0$ on the left and $x = 2$ on the right.

4. The region bounded by the curves $y = \sqrt{2x}$ and $y = x$.

5. The region bounded by the curves $y = \sqrt{8x}$ and $y = x^2$.

6. Do Exercise 2 a different way. (HINT: Use both positive and negative square roots.)

7. Do Exercise 4 a different way.

8. Do Exercise 5 a different way.

9. The region bounded by $y^2 = x$ and $y = x - 2$.

10. The region bounded by $x = y^2 - 1$ and $x = -y^2 + 1$.

In Exercises 11–22, find the volume of the solid of revolution formed when the indicated region is revolved about the given line. For each problem, sketch the graph of the region, draw a curved arrow around the axis of revolution, draw in a sample line segment, and state the name of the method you are using.

11. The region bounded above by $y = x^3$, below by the x-axis, and on the right by the vertical line $x = 2$, revolved about the x-axis.

12. The region of Exercise 11, revolved about the line $x = -1$.

13. The region of Exercise 11, revolved about the y-axis.

14. The region of Exercise 11, revolved about the line $x = 3$.

15. Do Exercise 11 by the other method.
16. Do Exercise 12 by the other method.
17. Do Exercise 13 by the other method.
18. Do Exercise 14 by the other method.
19. The region bounded above by $y = \sqrt{8x}$ and below by $y = x^2$, revolved about the x-axis.
20. The region of Exercise 19, revolved about the y-axis.
21. Do Exercise 19 by the other method.
22. Do Exercise 20 by the other method.

For Exercises 23 and 24, use the fact that the top half of a sphere of radius 2 is the solid of revolution formed when that part of the circle $x^2 + y^2 = 4$ in the first quadrant is revolved about the y-axis.

23. Use the above to find the volume of a hemisphere of radius 2, by the method of cross-sectional areas.
24. Use the above to find the volume of a hemisphere of radius 2, by the method of cylindrical shells.

The top half of a sphere of radius a is the solid of revolution formed when that part of the circle $x^2 + y^2 = a^2$ in the first quadrant is revolved about the y-axis. Use this to solve Exercises 25 and 26. You should imagine that you are *given* the first quadrant of a circle of radius a to revolve about the y-axis, so that you should think of a as a *constant*. This is important in taking antiderivatives correctly.

25. Use the above to find the volume of a hemisphere of radius a, by the method of cross-sectional areas.
26. Use the above to find the volume of a hemisphere of radius a, by the method of cylindrical shells.

12
MOTION

Sometimes you might be interested in getting the most precise information you can about a certain quantity you wish to study, so that you can predict how that quantity will behave over time. The quantity might be the position or velocity of a moving object, the concentration of medicine in a patient's bloodstream, the size of a pile of radioactive material, or the amount by which your savings account will grow. It is quite a remarkable fact that while there is not often an easily understood principle for describing the quantity you are interested in, there is an easily understood principle for describing *the rate of change* of that quantity. That is, there is a standard formula describing the rate of change, or derivative, of what you are interested in.

You should recognize by now that if you know the *derivative* of the function or quantity you are interested in, then you can just find an antiderivative to get the quantity. The only question is, which antiderivative? Remember, there are infinitely many of them, all differing by an arbitrary constant. What you need, in addition to the derivative of the quantity you are interested in, is the value of that quantity at one particular point.

The first section of this chapter deals with the purely mathematical formulation of the problem, and the second with one simple type of application, to motion. In this final short chapter, you will find only quite simple and specialized examples. Other applications, to more complicated types of motion, and to other phenomena as well, require more calculus (but not too much more) than has been dealt with in this book.

12.1 SIMPLE INITIAL VALUE PROBLEMS

An initial value problem is a problem in which you are given the derivative $\frac{dy}{dt}$ of an unknown function y, and the value y_0 of that function at some particular time t_0.

Example 12.1 Solve the initial value problem

$$\frac{dy}{dt} = t^2 + 3, \quad y(1) = 4$$

Solution: Since $\dfrac{dy}{dt} = t^2 + 3$,

$$y = \int (t^2 + 3)\, dt = \frac{1}{3}t^3 + 3t + C$$

Of the infinitely many antiderivatives of $t^2 + 3$, you want to find the one that has the value 4 when $t = 1$. Simply substitute 1 for t and 4 for y in the general antiderivative to get

$$4 = \frac{1}{3}\cdot 1^3 + 3\cdot 1 + C = \frac{10}{3} + C$$

Now solve for C, so that $C = \dfrac{2}{3}$. This is the particular value of C that will give you the solution of the initial value problem. The solution is thus

$$y = \frac{1}{3}t^3 + 3t + \frac{2}{3}$$

Example 12.2 Solve the initial value problem

$$\frac{dy}{dt} = t\sqrt{t^2 + 1}, \quad y(0) = 1$$

Solution: Since $\dfrac{dy}{dt} = t\sqrt{t^2 + 1}$,

$$y = \int t\sqrt{t^2 + 1}\, dt = \int t\left(t^2 + 1\right)^{\frac{1}{2}}\, dt$$

You can do this integral by substitution. Let $u = t^2 + 1$, so that $\dfrac{du}{dt} = 2t$. Thus $t\, dt = \dfrac{1}{2}du$, and

$$
\begin{aligned}
y = \int t\left(t^2 + 1\right)^{\frac{1}{2}}\, dt &= \int \frac{1}{2}u^{\frac{1}{2}}\, du \\
&= \frac{1}{2}\cdot\frac{2}{3}u^{\frac{3}{2}} \\
&= \frac{1}{3}\left(t^2 + 1\right)^{\frac{3}{2}} + C
\end{aligned}
$$

Substituting $t = 0$ and $y = 1$ into the above general antiderivative gives

$$1 = \frac{1}{3}\cdot 1^{\frac{3}{2}} + C = \frac{1}{3} + C$$

So, $C = \dfrac{2}{3}$, and the solution to the initial value problem is

$$y = \frac{1}{3}\left(t^2 + 1\right)^{\frac{3}{2}} + \frac{2}{3}$$

12.2 MOTION WITH CONSTANT ACCELERATION

There is one point you should be aware of in applying the methods of calculus to real world problems of motion. To apply the methods of calculus, you must be able to think precisely about the specific position of an object at a specific time. However, an apple falls from a tree, a rocket is launched up into the air, a car comes to a stop, and a thrown ball rises and then falls, without any of us necessarily around to time the event with a stopwatch or to measure positions with a ruler.

The first step in dealing with these problems of motion by mathematical methods is to apply your own stopwatch and ruler to the problem. This basically means deciding upon how you should measure time and position, how you should set up an axis for time and an axis for position. What's involved in setting up an axis? Basically three things:

1. choosing which direction should be positive and which should be negative;
2. picking an origin (deciding which point should correspond to 0);
3. choosing a unit.

Usually, the words of the problem will define natural units. If the problem talks about feet per second, then a foot will be the unit of position (or length) and a second the unit of time. If a problem talks about miles per hour, then a mile will be the unit of position, and an hour the unit of time.

How about picking an origin, and a positive direction? Obviously, it should not matter how you do this. The apple will fall as it falls, regardless of whether you consider the position of the apple to be its height above the ground (which is essentially calling ground level position 0, and up as the positive direction), or whether you consider the position of the apple to be measured by how far it has fallen from its original position (which is essentially calling the apple's original position the origin, and specifying that down is the positive direction). Likewise, the apple will fall as it falls, regardless of whether you say that it started falling at 2:30 P.M. (which is sort of like declaring noon to be the origin of the time axis), or whether you say that it started falling at time $t = 0$ (which is declaring the origin of the time axis to be the instant the apple starts to fall). You can set up the axes any way you want, but once you set them up you must be consistent. The best way to ensure consistency is to be aware that before you work any problem, you really must set up axes, and to do so consciously.

You must decide which instant of time during the course of the problem should be considered as time $t = 0$. Choosing the positive direction for time isn't so difficult—you're not likely to see a problem that doesn't consider the future to be the positive direction for time, and the past to be the negative direction. Often the choice of time $t = 0$ is fairly straightforward also. If you throw a ball or shoot off a rocket, a natural choice for $t = 0$ is the instant the object starts to move.

You must also decide which position should be position 0. For a ball being thrown down from the top of a building, there are two natural choices: the top of the building or ground level. In order to try to deal with all problems of vertical motion consistently, in this chapter I will always choose ground level as position 0, and up as the positive direction.

Suppose you know that an object is moving in a straight line, with constant acceleration. The definition of **acceleration** is *the rate of change of velocity with respect to time*. Using the letter v to denote the velocity function, and the letter a to denote acceleration, you have that

$$\frac{dv}{dt} = a$$

and that a is a constant. It follows that

$$v(t) = \int \frac{dv}{dt}\, dt = \int a\, dt = at + C$$

To obtain the particular velocity function for a particular moving object, you must know the velocity at some particular instant of time. Often problems will give you the *initial velocity* of the object, that is, the velocity at a time that is natural to call time $t = 0$. Using the symbol v_0 to stand for initial velocity, and substituting 0 in for t and v_0 in for v yields

$$v_0 = a \cdot 0 + C = C$$

Thus, the particular velocity function for the problem at hand is

$$v(t) = at + v_0$$

Once you know the velocity function, you know the derivative with respect to time of the position function. Recall from Section 2.2 that velocity means the rate of change of position with respect to time. Using the letter y to denote the position function, you have that

$$\frac{dy}{dt} = v(t) = at + v_0$$

Thus,

$$y = \int \frac{dy}{dt}\, dt = \int (at + v_0)\, dt = \frac{1}{2}at^2 + v_0 t + C$$

As before, to find the particular position function for the problem at hand, you must be given the particular position of the object at some particular time. Often (but certainly not always) this is given as the position of the object at a time that is natural to call time $t = 0$. Using the symbol y_0 to stand for this initial position, and substituting 0 for t and y_0 for y yields

$$y_0 = \frac{1}{2}a \cdot 0^2 + v_0 \cdot 0 + C = C$$

So, the particular position function for the problem at hand is

$$y(t) = \frac{1}{2}at^2 + v_0 t + y_0$$

Note to students: You should always be aware that the above formulas for velocity and position hold *only when* the acceleration is constant, and the particular given values of velocity and position are the values at time $t = 0$. Rather than using these formulas in word problems involving motion, you should always solve such problems by modeling what I've done above. If you know the acceleration $a = \dfrac{dv}{dt}$, and the value of the velocity v at some particular time, then you can solve an initial value problem to get the velocity function $v(t)$. If you know the position at some particular time, then since you know $v = \dfrac{dy}{dt}$, you can solve an initial value problem to get the position function $y(t)$.

For objects moving vertically near the surface of the earth, the acceleration due to the force of gravity is nearly constant. The letter g is used to denote the value of this acceleration, and g is approximately equal to 32 feet/second². If you adopt the convention that up is always positive, then since the acceleration of gravity acts in a downward fashion, you should use $a = -g = -32$ feet/second².

Example 12.3 Juanita is standing on the roof of a building 192 feet tall, and throws a ball up into the air with an initial speed of 64 feet/second. (a) Find the formulas for the velocity and the position of the ball at a later time t. (b) How high does the ball go? (c) When does the ball hit the ground?

Solution:

(a) As mentioned before, I will adopt the convention that ground level is position $y = 0$, and that up is positive. I will also consider time $t = 0$ to be the instant Juanita throws the ball. The ball has a constant acceleration due to gravity of -32 ft/s², so that $a = -32$. Thus $\dfrac{dv}{dt} = -32$, and

$$v(t) = \int -32 \, dt = -32t + C$$

Recall that speed is the absolute value of velocity. Thus, while speed is always positive, velocity is positive or negative, depending upon whether the object is traveling in the positive direction or in the negative direction. As Juanita initially throws the ball *up*, she is throwing the ball in the positive direction, and the ball has initial velocity $v(0) = 64$. Substituting 0 for t and 64 for v gives

$$64 = -32 \cdot 0 + C = C$$

so that

$$v(t) = -32t + 64$$

As $\dfrac{dy}{dt} = v(t)$,

$$y(t) = \int \frac{dy}{dt}\, dt = \int (-32t + 64)\, dt = -16t^2 + 64t + C$$

Counting ground level as position $y = 0$, the initial position is the height of the building, so that $y(0) = 192$. Substituting 0 for t and 192 for y gives

$$192 = -16 \cdot 0^2 + 64 \cdot 0 + C = C$$

Thus,

$$y(t) = -16t^2 + 64t + 192$$

(b) Recall from Section 2.3 that the ball reaches its high point at the instant when its velocity equals 0. Since you know the velocity function, you can determine *when* the ball reaches its high point, and then substitute that value of t into the position function to determine the high point. The hard part of solving this part of the example is for you to know that you must first determine *when* the ball reaches its high point. As $v(t) = -32t + 64 = 0$ at time $t = 2$, the ball reaches its high point after 2 seconds, and its position then is

$$y(2) = -16 \cdot 2^2 + 64 \cdot 2 + 192 = -64 + 128 + 192 = 256 \text{ feet}$$

(c) The ball hits the ground when it is at position $y = 0$. To find out when that is, simply set the formula for $y(t)$ equal to 0, and solve for t.

$$y(t) = -16t^2 + 64t + 192 = -16(t^2 - 4t - 12) = -16(t - 6)(t + 2)$$

Clearly time $t = -2$ is before Juanita threw the ball, and is not relevant for the problem. Thus the ball hits the ground after 6 seconds.

If the above example had read that Juanita throws the ball *down* with an initial speed of 64 feet/second, the only change in the above work would be that the initial velocity would be $v_0 = v(0) = -64$ feet/second. Of course, this would give completely different answers to parts (b) and (c). In fact, part (b) would really no longer make sense, but it is clear that the ball would hit the ground in less than 6 seconds. If the above example had read that Juanita *drops* the ball, then the initial velocity would be $v_0 = 0$. Dropping the ball just means letting it go, without throwing it either up or down.

Example 12.4 Fred is standing on the ground, and throws a ball up into the air. He observes that it falls back to the ground 5 seconds later. What was the initial velocity of the ball?

Solution: As in the previous problem, ground level will be position 0, up will be the positive direction, and time $t = 0$ will be the instant when Fred first throws the ball. The acceleration due to gravity will be $a = -32$. Thus $\dfrac{dv}{dt} = -32$, so that $v(t) = -32t + C$. The problem does not give any information about the particular velocity of the ball at any particular time. In fact, the problem is to find the initial velocity. All you can do at this stage is to use the symbol v_0 for the initial velocity, at time $t = 0$, to get $v(t) = -32t + v_0$, and keep on going. As $v(t) = \dfrac{dy}{dt}$,

$$y(t) = \int (-32t + v_0)\, dt = -16t^2 + v_0 t + C$$

Since Fred was standing on the ground when he first threw the ball, the initial position is ground level, $y = 0$. Substituting 0 for t and 0 for y gives $0 = -16 \cdot 0^2 + v_0 \cdot 0 + C = C$, so that $C = 0$ and

$$y(t) = -16t^2 + v_0 t$$

The key to solving the problem is to notice that you know the position of the ball at the later time $t = 5$. It is back on the ground, and so $y(5) = 0$. Substituting 5 for t and 0 for y into the formula for $y(t)$ gives

$$0 = -16 \cdot 5^2 + v_0 \cdot 5 = -400 + 5v_0$$

This can easily be solved to give $v_0 = 80$ ft/s.

Example 12.5 A driver traveling at the speed of 60 miles per hour (which equals 88 feet per second) slams on his brakes. Assuming that the brakes provide a constant deceleration of 22 feet/second², how long does it take the driver to come to a stop?

Solution: In this problem, it seems reasonable to use the letter x rather than y for position, since the car is traveling horizontally (more or less). It also seems reasonable to choose $x = 0$ to be the position of the car when the driver first applies his brakes, and the positive direction to be the direction the car is traveling in. Furthermore, it seems reasonable to call time $t = 0$ the instant the driver slams on his brakes. The word "deceleration" means negative acceleration (you decelerate when you slow down), so that $a = -22$. It follows that

$$v(t) = \int -22 \, dt = -22t + C$$

You are given that at time $t = 0$ the velocity is 88 ft/s, so substitute 0 for t and 88 for v in the above equation to get $C = 88$, and $v(t) = -22t + 88$. It follows that

$$x(t) = \int (-22t + 88) \, dt = -11t^2 + 88t + C$$

The axes were set up so that at $t = 0$, you have $x = 0$ also. Substituting 0 for both t and x yields $x(t) = -11t^2 + 88t$. If you knew when the car came to a stop, you could substitute that value for t into the equation for $x(t)$ and get the answer. But you do know when the car comes to a stop—when its velocity equals 0! Set $v(t) = -22t + 88$ equal to 0, and solve for t, to get $t = 4$. The position of the car at time $t = 4$ is

$$x(4) = -11 \cdot 4^2 + 88 \cdot 4 = -176 + 352 = 176$$

The car skids for 176 feet before coming to a stop.

12.3 SUMMARY OF MAIN POINTS

- In an initial value problem, you are given the derivative of an unknown function, as well as the value of that function at some particular point.

- In a simple initial value problem, the formula for the derivative involves only the variable for the function, and solving the initial value problem involves first finding a general antiderivative.

- Once you have a general antiderivative, substitute in the values you are given (the initial values) for the variable and the function, and solve for the arbitrary constant C to get the solution of the initial value problem.

- Initial value problems can be used to find the velocity and then the position of an object moving in a straight line, when the acceleration of the object is known.

- The acceleration due to gravity near the surface of the earth is nearly constant, and is equal to -32 ft/s^2 (if up is the positive direction).

- Before you can apply mathematical techniques to problems involving straight line motion, you must set up axes for the measurement of position and time.

- Setting up an axis means choosing a unit, an origin, and a positive direction.

12.4 EXERCISES

In Exercises 1–3, solve the initial value problems.

1. $\dfrac{dy}{dt} = \dfrac{1}{\sqrt{t}}, \quad y(4) = 3$

2. $\dfrac{dx}{dt} = t^3 - 3t, \quad x(0) = 1$

3. $\dfrac{dy}{dx} = 4x + \dfrac{1}{x^2}, \quad y(2) = 0$

Solve the problems of Exercises 4–8. For those involving vertical motion, assume constant acceleration due to gravity of -32 feet/second^2.

4. Juanita is standing on the roof of a building 192 feet tall, and throws a ball down with an initial speed of 64 feet/second. (a) Find the formulas for the velocity and the position of the ball at a later time t. (b) When does the ball hit the ground?

5. Juanita is standing on the roof of a building 192 feet tall, and drops a ball. (a) Find the formulas for the velocity and the position of the ball at a later time t. (b) When does the ball hit the ground?

6. A ball thrown up into the air with a speed of 64 feet/second from the roof of a building hits the ground in 7 seconds. How tall is the building?

7. A ball thrown from the roof of a 224-foot-tall building hits the ground in 8 seconds. What was the initial velocity of the ball?

8. A motorist slams on her brakes, and skids for 224 feet. Assuming that the brakes provide a constant deceleration of 28 feet/second^2, what was the motorist's initial speed when she slammed on the brakes?

APPENDICES

In this book, I have tried to provide a leisurely, conceptual approach to the main ideas of calculus—the ideas of the derivative and of the integral—while at the same time providing an introduction to some of the basic techniques and applications of calculus. My foremost hope is that you understand the main idea of differential calculus—what an instantaneous rate of change is; and the main ideas of integral calculus—how some problems can be solved by an estimation procedure involving breaking the problem up into small pieces on each of which you can approximate a solution; and how the Fundamental Theorem of Calculus and the use of antiderivative provides an alternate, easier computational technique. I hope that the symbols and techniques of calculus have conceptual, verbal, and geometric *meaning* for you.

If I tried to cover in the same fashion all of the usual topics of a first-year calculus course, the book would be so heavy you would get a hernia trying to lug it around. However, for the convenience of students who feel that I am omitting too many of the topics they are interested in, this appendix contains a brief summary of extra material involving the trigonometric, exponential, and logarithmic functions. It is in the nature of an expanded "Summary of Main Points," rather than a conceptual explanation.

A
THE TRIGONOMETRIC FUNCTIONS

A.1 DEFINITIONS

Just as length can be measured in different units, such as feet and meters, angles also can be measured in different units. The two standard units for measuring angles are **degrees** and **radians**. An angle of 5 degrees is written as $5°$, while an angle of 5 radians is written as 5 rad or simply 5. The mathematical default for specifying angles is in radians, so that if units are not mentioned, assume the angle is in radians. Learn how to use your calculator in both radian mode and degree mode.

To convert between radians and degrees, you just have to remember the measure of one angle in both units. An angle of 180 degrees is the same as an angle of π radians:

$$180° = \pi \text{ rad} \tag{A.1}$$

From this it follows that an angle of

$$1° = \frac{1}{180} 180° = \frac{1}{180} \pi \text{ rad} = \frac{\pi}{180} \text{ rad}$$

So, for example,

$$37° = 37 \cdot 1° = 37 \cdot \frac{\pi}{180} = \frac{37\pi}{180} \text{ rad}$$

It also follows that an angle of

$$1 \text{ rad} = \frac{1}{\pi} \pi \text{ rad} = \frac{1}{\pi} 180° = \left(\frac{180}{\pi}\right)°$$

Thus, for example,

$$3 \text{ rad} = 3 \cdot \left(\frac{180}{\pi}\right)° = \left(\frac{540}{\pi}\right)°$$

and

$$\frac{\pi}{6} \text{ rad} = \frac{\pi}{6} \left(\frac{180}{\pi}\right)° = 30°$$

The two most important trigonometric (trig) functions are the sine, abbreviated sin, and the cosine, abbreviated cos. Let x be an angle (not the right angle) in a right triangle, and label the leg of the triangle adjacent to x as A, the leg opposite angle x as O, and the hypotenuse as H, as pictured in Figure A.1.

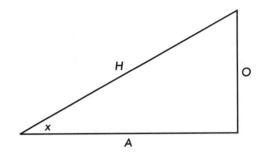

Figure A.1

Then

$$\sin x = \frac{O}{H} \tag{A.2}$$

$$\cos x = \frac{A}{H} \tag{A.3}$$

For angles not between $0°$ and $90°$, that is, not between 0 and $\frac{\pi}{2}$, the unit circle picture is helpful (see Figure A.2). Draw a unit circle centered at the origin. The standard way to draw an angle x (rad) is to have the vertex of the angle at the center of the circle, and one ray of the angle along the positive half of the horizontal axis. The convention for drawing the other ray of the angle is that if the angle $x > 0$, the angle is swept out in the counterclockwise direction, while if $x < 0$ the angle is swept out in the clockwise direction. The second ray of the angle cuts the circle at a point $P = (a, b)$. With this convention,

$$\sin x = b \tag{A.4}$$

while

$$\cos x = a \tag{A.5}$$

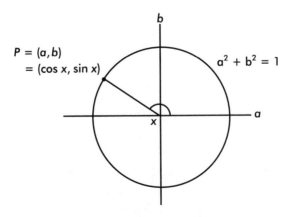

Figure A.2

As every point $P = (a, b)$ on the unit circle centered at the origin satisfies the equation $a^2 + b^2 = 1$, you have the fundamental identity

$$\sin^2 x + \cos^2 x = 1, \text{ for all } x \qquad (A.6)$$

The other four trig functions are tangent (abbreviated tan), cotangent (abbreviated cot), secant (abbreviated sec) and cosecant (abbreviated csc). They can all be defined in terms of sin and cos, and thus for angles x with $0 < x < \frac{\pi}{2}$, these trig functions can also be defined in terms of ratios of sides of a right triangle, just as sin and cos were.

$$\tan x \; = \; \frac{\sin x}{\cos x} = \frac{O}{A} \qquad (A.7)$$

$$\cot x \; = \; \frac{\cos x}{\sin x} = \frac{1}{\tan x} = \frac{A}{O} \qquad (A.8)$$

$$\sec x \; = \; \frac{1}{\cos x} = \frac{H}{A} \qquad (A.9)$$

$$\csc x \; = \; \frac{1}{\sin x} = \frac{H}{O} \qquad (A.10)$$

Of course, since the above four trig functions are defined by fractional expressions, they are each undefined for certain values of x, depending upon whether or not $\sin x$ or $\cos x = 0$.

You should learn the ratios of the sides of two standard triangles. The ratios of sides of a $30°, 60°, 90°$ triangle are $1, 2, \sqrt{3}$. As 2 is the largest of these numbers, it is the hypotenuse, while the smallest of the numbers, 1, is opposite the smallest angle, the $30°$ angle. In a $45°, 45°, 90°$ right triangle, the ratios are $1, 1, \sqrt{2}$. The two triangles are pictured in Figure A.3.

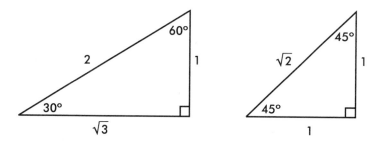

Figure A.3

Using the above triangles and the circle picture of Figure A.2, you should be able to figure out the values of the trig functions at certain standard angles.

A.2 IDENTITIES

Upon dividing both sides of the fundamental equation $1 = \sin^2 x + \cos^2 x$ by $\sin^2 x$, and using the definitions of the other four trig functions, you get

$$\csc^2 x = 1 + \cot^2 x \qquad\qquad \text{(A.11)}$$

Similarly, dividing both sides of the fundamental equation by $\cos^2 x$ gives

$$\sec^2 x = \tan^2 x + 1 \qquad\qquad \text{(A.12)}$$

There are addition formulas for sin and cos, whose proofs you can find in any precalculus book. They are

$$\sin(x + y) = \sin x \cos y + \cos x \sin y \qquad\qquad \text{(A.13)}$$

and

$$\cos(x + y) = \cos x \cos y - \sin x \sin y \qquad\qquad \text{(A.14)}$$

If you let $y = x$ in the above two formulas, you get the double-angle formulas

$$\sin(2x) = 2 \sin x \cos x \qquad\qquad \text{(A.15)}$$

and

$$\cos(2x) = \cos^2 x - \sin^2 x \qquad\qquad \text{(A.16)}$$

Equation A.16 can be used together with Equation A.6 to give identities for $\cos^2 x$ and for $\sin^2 x$, as follows:

$$
\begin{aligned}
\cos(2x) &= \cos^2 x - \sin^2 x \\
&= (1 - \sin^2 x) - \sin^2 x \\
&= 1 - 2 \sin^2 x
\end{aligned}
$$

So, $2 \sin^2 x = 1 - \cos(2x)$, and

$$\sin^2 x = \frac{1 - \cos(2x)}{2} \qquad\qquad \text{(A.17)}$$

Similarly,

$$
\begin{aligned}
\cos(2x) &= \cos^2 x - \sin^2 x \\
&= \cos^2 x - (1 - \cos^2 x) \\
&= 2 \cos^2 x - 1
\end{aligned}
$$

So, $2 \cos^2 x = 1 + \cos(2x)$, and

$$\cos^2 x = \frac{1 + \cos(2x)}{2} \qquad\qquad \text{(A.18)}$$

A.3 DERIVATIVES

Recall that the derivative $f'(x)$ of a function $f(x)$ is defined by

$$f'(x) = \lim_{h \to 0} \frac{f(x+h) - f(x)}{h}$$

Applying this to $f(x) = \sin x$ gives

$$
\begin{aligned}
(\sin x)' &= \lim_{h \to 0} \frac{\sin(x+h) - \sin x}{h} \\
&= \lim_{h \to 0} \frac{(\sin x \cos h + \cos x \sin h) - \sin x}{h} \quad \text{(Eq. A.13)} \\
&= \sin x \lim_{h \to 0} \left(\frac{\cos h - 1}{h} \right) + \cos x \lim_{h \to 0} \left(\frac{\sin h}{h} \right) \quad \text{(A.19)}
\end{aligned}
$$

If you knew the limits

$$\lim_{h \to 0} \left(\frac{\cos h - 1}{h} \right) \text{ and } \lim_{h \to 0} \left(\frac{\sin h}{h} \right)$$

you would know the derivative of $\sin x$. Your regular calculus text probably has a geometric proof of what these limits are, but why don't you try using your calculator to see if you can experimentally determine these limits? Evaluate both $\frac{\cos h - 1}{h}$ and $\frac{\sin h}{h}$ for values of h getting closer and closer to 0, such as $h = 1, 0.1, 0.001, 0.0001$, and see what you get. (Note: Be sure your calculator is in radian mode before you do this.) You should get

$$\lim_{h \to 0} \left(\frac{\cos h - 1}{h} \right) = 0 \quad \text{(A.20)}$$

and

$$\lim_{h \to 0} \left(\frac{\sin h}{h} \right) = 1 \quad \text{(A.21)}$$

Using these two limits and Equation A.19, you get that

$$(\sin x)' = \cos x \quad \text{(A.22)}$$

Similarly, using the definition of the derivative, the addition rule for cosines (Equation A.14), and the above two limits, you can deduce that

$$(\cos x)' = -\sin x \quad \text{(A.23)}$$

The derivatives of the other four trig functions can be found by writing them in terms of sines and cosines. For example, $\tan x = \dfrac{\sin x}{\cos x}$, so by applying the quotient rule and the formulas for the derivative of sine and cosine, you get

$$
\begin{aligned}
(\tan x)' &= \left(\frac{\sin x}{\cos x}\right)' \\
&= \frac{\cos x \cdot (\sin x)' - \sin x \cdot (\cos x)'}{\cos^2 x} \\
&= \frac{\cos x \cdot \cos x - \sin x \cdot (-\sin x)}{\cos^2 x} \\
&= \frac{\cos^2 x + \sin^2 x}{\cos^2 x} \\
&= \frac{1}{\cos^2 x} \\
&= \sec^2 x
\end{aligned}
$$

Thus,

$$(\tan x)' = \sec^2 x \tag{A.24}$$

Similarly,

$$
\begin{aligned}
(\cot x)' &= -\csc^2 x & \text{(A.25)} \\
(\sec x)' &= \sec x \tan x & \text{(A.26)} \\
(\csc x)' &= -\csc x \cot x & \text{(A.27)}
\end{aligned}
$$

Example A.1 Find the derivative of $x^3 \tan x$.

Solution: Use the product rule. Thus,

$$
\begin{aligned}
(x^3 \tan x)' &= x^3 (\tan x)' + (x^3)' \tan x \\
&= x^3 \sec^2 x + 3x^2 \tan x
\end{aligned}
$$

Example A.2 Find the derivative of $\sec 3x$.

Solution: The function $\sec 3x$ is a composite function, with the secant function on the outside, and the function $3x$ on the inside. Thus you have to use the chain rule. You might want to rewrite $\sec 3x$ with parentheses, as $\sec(3x)$, to emphasize the composite nature of the function, and as a reminder to yourself to use the chain rule. Let $u = 3x$. Then $y = \sec(3x) = \sec u$ and

$$
\begin{aligned}
\frac{dy}{dx} &= \frac{dy}{du}\frac{du}{dx} \\
&= \sec u \tan u \cdot 3 \\
&= \sec 3x \tan 3x \cdot 3 \\
&= 3 \sec 3x \tan 3x
\end{aligned}
$$

Example A.3 Find the derivative of $y = \sin^3 x$.

Solution: The function y is also a composite function, and you might want to rewrite it to emphasize that fact more clearly. Thus $y = (\sin x)^3$. Let $u = \sin x$. Then $y = (\sin x)^3 = u^3$, and, by the chain rule,

$$
\begin{aligned}
\frac{dy}{dx} &= \frac{dy}{du}\frac{du}{dx} \\
&= 3u^2 \cos x \\
&= 3\sin^2 x \cos x
\end{aligned}
$$

Example A.4 Find the derivative of $y = \cot^4 3x$.

Solution: This function is a composite of three different functions, so you have to use the chain rule twice. First rewrite the function so that it is clearer how the three different functions are composed. Thus,

$$ y = \cot^4 3x = [\cot(3x)]^4 $$

Applying the chain rule one step at a time gives

$$
\begin{aligned}
y' &= 4[\cot(3x)]^3 \cdot (\cot(3x))' \\
&= 4[\cot(3x)]^3 \cdot \left(-\csc^2(3x)\right) \cdot (3x)' \\
&= 4[\cot(3x)]^3 \cdot \left(-\csc^2(3x)\right) \cdot 3 \\
&= -12\cot^3 3x \csc^2 3x
\end{aligned}
$$

A.4 ANTIDERIVATIVES

Every time you learn a new derivative formula, you have at the same time learned a new integral or antiderivative formula. As $(\sin x)' = \cos x$, it is immediate that $\int \cos x \, dx = \sin x + C$, and that, by the Fundamental Theorem of Calculus,

$$ \int_0^{\frac{\pi}{2}} \cos x \, dx = \sin x \Big|_0^{\frac{\pi}{2}} = \sin\left(\frac{\pi}{2}\right) - \sin 0 = 1 - 0 = 1 $$

$$ \int \sin x \, dx = -\cos x + C \tag{A.28} $$

$$\int \cos x \, dx = \sin x + C \qquad \text{(A.29)}$$

$$\int \sec^2 x \, dx = \tan x + C \qquad \text{(A.30)}$$

$$\int \csc^2 x \, dx = -\cot x + C \qquad \text{(A.31)}$$

$$\int \sec x \tan x \, dx = \sec x + C \qquad \text{(A.32)}$$

$$\int \csc x \cot x \, dx = -\csc x + C \qquad \text{(A.33)}$$

Below are three examples showing how the above formulas can be used, together with simple substitution.

Example A.5 Evaluate $\int \sin^3 x \cos x \, dx$.

Solution: This problem has exactly the same pattern as an integral like $\int (x^6 + 1)^3 6x^5 \, dx$. You are integrating a power of a function of x, multiplied by the derivative of the function being raised to a power. If you let u be the function being raised to a power, you will get an integral of the form $\int u^n \, du$.

Thus let $u = \sin x$, so that $\dfrac{du}{dx} = \cos x$, and $du = \cos x \, dx$. It follows that

$$\begin{aligned}
\int \sin^3 x \cos x \, dx &= \int u^3 \, du \\
&= \frac{1}{4} u^4 \\
&= \frac{1}{4} \sin^4 x + C
\end{aligned}$$

Example A.6 Evaluate $\int x \cos(x^2) \, dx$.

Solution: Here the inside part of the composite is x^2, and since you have (almost) the derivative of the inside part as a factor in the integrand, substitution of $u = x^2$ will do the trick. So let $u = x^2$. Then $\dfrac{du}{dx} = 2x$, so that $du = 2x \, dx$ and $x \, dx = \dfrac{1}{2} du$. With this substitution,

$$\int x \cos(x^2) \, dx = \int \frac{1}{2} \cos u \, du$$

$$= \frac{1}{2} \sin u$$

$$= \frac{1}{2} \sin(x^2) + C$$

Example A.7 Evaluate $\displaystyle\int \sec 3x \tan 3x \, dx$.

Solution: You should recognize $\sec \cdot \tan$ as the derivative of sec. The only problem is how to handle the $3x$. Since the derivative of $3x$ is a constant, and multiplicative constants can be moved outside of the integral sign, letting $u = 3x$ is the right substitution. With this substitution, $\dfrac{du}{dx} = 3$, so that $du = 3 \, dx$ and $dx = \dfrac{1}{3} du$. Thus,

$$\int \sec 3x \tan 3x \, dx = \int \frac{1}{3} \sec u \tan u \, du$$

$$= \frac{1}{3} \int \sec u \tan u \, du$$

$$= \frac{1}{3} \sec u$$

$$= \frac{1}{3} \sec 3x + C$$

B

EXPONENTIAL AND LOGARITHMIC FUNCTIONS

B.1 DEFINITIONS

An **exponential** function is a function of the form $f(x) = a^x$, where the number a, called the *base*, is a positive constant. If x is a positive integer n, then a^n is simply a multiplied by itself n times, so that, for example, $a^4 = a \cdot a \cdot a \cdot a$. If x is a unit fraction, that is, a fraction with 1 in the numerator, such as $\frac{1}{n}$ (n a positive integer), then $a^{\frac{1}{n}}$ is the nth root of a, that is, the number whose nth power equals a. Thus $16^{\frac{1}{4}}$ is the fourth root of 16, the number whose fourth power equals 16, namely, 2. If x is a positive rational number of the form $\frac{p}{q}$, where p and q are positive integers, then

$$a^{\frac{p}{q}} = (a^p)^{\frac{1}{q}} = \left(a^{\frac{1}{q}}\right)^p.$$

It is usually easier to compute the root first, since that will keep the numbers you are dealing with smaller. Thus,

$$16^{\frac{3}{4}} = \left(16^{\frac{1}{4}}\right)^3 = 2^3 = 8.$$

If you imagined plotting points of the graph of $f(x) = a^x$ for all positive rational numbers x, you would have a lot of points plotted (in fact, infinitely many). The easiest and most intuitive way of explaining how to fill in the rest of the graph is to imagine just connecting the points you have already plotted, as smoothly as possible. The values of the exponential function on the negative half-line are determined by the equation $a^{-x} = \frac{1}{a^x}$.

An **inverse function** g to a function f is a function that undoes what f does. Thus if f takes the number a and assigns it a value of b, then g will take the number b and assign it a value of a. That is,

$$f(a) = b \text{ means precisely the same as } g(b) = a$$

Inverse functions do not always exist. For example, if you tried to define an inverse g to the function $f(x) = x^2$, then $g(4)$ would be the number whose square is 4. There are *two* such numbers, 2 and –2, and thus $g(4)$ would be either 2 or –2. Such ambiguity is not allowed in the definition of a function— part of the definition is that a function should have a unique value at each point. Of course, for the squaring function, the dilemma can be resolved by considering the squaring function as a function defined not on the whole real line, but as a function defined only for $x \geq 0$. Then $g(4)$ would be the number $x \geq 0$ whose square is 4, and there is only one such number, namely 2. This inverse function $g(x)$ is precisely $\sqrt{x}$, where you will recall that $\sqrt{x}$ means the positive square root of x.

If $a > 1$, then the function $f(x) = a^x$ is strictly increasing, and it does have an inverse. Inverses of exponential functions are called **logarithmic** functions, and the inverse of $f(x) = a^x$ is $g(x) = \log_a x$, read as "log to the base a of x." You can think of $\log_a x$ as the power you must raise a to to get x just like you can think of $\sqrt{x}$ as the (positive) number you must square to get x.

There is a special number e, which has a value of about 2.718. It is not too easy to explain exactly where e comes from. Of all the possible different functions of the form $f(x) = a^x$, for any positive number a, the one whose derivative has the simplest formula is the function $f(x) = e^x$. Why should this be so? Well, remember that the derivative of a function is defined in terms of a limit. The number e can also be defined in terms of a limit, and it is the relationship between the definition of e in terms of a limit, and the limit that appears in working out the derivative of an exponential function, that makes the derivative formula the simplest when the base is e.

The function $\log_{10} x$ is called the *common logarithm*, and is often written simply as $\log x$. The function $\log_e x$ is called the *natural logarithm*, and is written as $\ln x$.

B.2 IDENTITIES

An exponential function $f(x) = a^x$ satisfies the following:

$$a^x a^y = a^{x+y} \tag{B.1}$$

$$(a^x)^y = a^{xy} \tag{B.2}$$

$$a^0 = 1 \tag{B.3}$$

$$\frac{a^x}{a^y} = a^{x-y} \tag{B.4}$$

$$a^{-x} = \frac{1}{a^x} \tag{B.5}$$

The first two of the above equations are clear when x and y are positive integers.

For $a > 1$, the relationships between the exponential function a^x and its inverse function $\log_a x$ can be summarized as follows:

$$a^x = y \text{ means precisely the same as } \log_a y = x \tag{B.6}$$

$$a^{\log_a x} = x \tag{B.7}$$

$$\log_a(a^x) = x \tag{B.8}$$

You should not think of these equations as difficult or mysterious. They are trivial, provided you think about logarithmic functions in the right way. If you think of $\log_a y$ as the power you must raise a to to get y, then the above equations become as easy to understand as the answer to the question "Who is buried in Grant's tomb?"

If $\log_a y$ is the power you must raise a to to get y, and $a^x = y$, then clearly x is the power you must raise a to to get y, and thus $\log_a y = x$. For the second of the above equations, think of $\log_a x$ as the power you must raise a to to get x. Then if you raise a to that power, what else can you get but x? The left-hand side of the third of the above equations can be thought of as asking what power you must raise a to to get a^x. The answer is clearly x.

The following rules for logarithms follow from the definition of the logarithm as the inverse of the exponential function, and the rules given earlier for exponentials.

$$\log_a(xy) = \log_a x + \log_a y \tag{B.9}$$
$$\log_a(x^y) = y \log_a x \tag{B.10}$$
$$\log_a 1 = 0 \tag{B.11}$$
$$\log_a\left(\frac{x}{y}\right) = \log_a x - \log_a y \tag{B.12}$$
$$\log_a\left(\frac{1}{y}\right) = -\log_a y \tag{B.13}$$

B.3 DERIVATIVES

Let $f(x) = a^x$. Using the definition of the derivative, and the rules for exponential functions, it follows that

$$
\begin{aligned}
(a^x)' &= \lim_{h \to 0} \frac{a^{x+h} - a^x}{h} \\
&= \lim_{h \to 0} \frac{a^x a^h - a^x}{h} \\
&= a^x \cdot \left(\lim_{h \to 0} \frac{a^h - 1}{h} \right)
\end{aligned}
$$

It turns out that if you choose a to be the number e, then

$$\lim_{h \to 0} \left(\frac{e^h - 1}{h} \right) = 1$$

This means from the above calculations that

$$(e^x)' = e^x \tag{B.14}$$

If $y = e^u$ where u is a function of x, then by the chain rule,

$$\frac{d}{dx}\left(e^u\right) = \frac{d}{du}\left(e^u\right)\frac{du}{dx} = e^u\frac{du}{dx} \tag{B.15}$$

Example B.1 Find the derivative of $y = e^{\sin x}$.

> **Solution:** The function $e^{\sin x}$ is a composite with the exponential function on the outside, and the function $\sin x$ on the inside. The part in the exponent can be considered the inside part. Thus let $u = \sin x$, so that $y = e^u$. Then
>
> $$\begin{aligned}\frac{dy}{dx} &= \frac{dy}{du}\frac{du}{dx} \\ &= e^u \cos x \\ &= e^{\sin x} \cos x\end{aligned}$$

Knowing the derivative of e^x allows you to find the derivative of the inverse function $\ln x$ as follows: By the relationship between the exponential function and its inverse, the natural logarithm function,

$$e^{\ln x} = x$$

Take the derivative with respect to x of both sides. For the left-hand side, you must use the chain rule, with $\ln x$ as the inside part. For the part of the chain rule involving the derivative of the inside part, you can only write the *symbol* for the derivative of the inside part $\ln x$, because you do not know it yet. Thus,

$$\begin{aligned}\left[e^{\ln x}\right]' &= (x)' \\ e^{\ln x} \cdot (\ln x)' &= 1 \\ (\ln x)' &= \frac{1}{e^{\ln x}} \\ (\ln x)' &= \frac{1}{x} \tag{B.16}\end{aligned}$$

It follows by the chain rule that if $y = \ln u$, where u is a function of x, then

$$\frac{d}{dx}\left(\ln u\right) = \frac{d}{du}\left(\ln u\right)\frac{du}{dx} = \frac{1}{u}\frac{du}{dx} \tag{B.17}$$

Example B.2 Find the derivative of $y = \tan(\ln x)$.

> **Solution:** $y = \tan u$, where $u = \ln x$. Thus by the chain rule,
>
> $$\frac{dy}{dx} = \frac{dy}{du}\frac{du}{dx}$$

$$= \sec^2 u \cdot \frac{1}{x}$$

$$= \sec^2(\ln x) \cdot \frac{1}{x}$$

$$= \frac{\sec^2(\ln x)}{x}$$

Example B.3 Find the derivative of $y = \ln(\tan x)$.

Solution: Here $y = \ln u$, with $u = \tan x$, so by the chain rule

$$\frac{dy}{dx} = \frac{dy}{du}\frac{du}{dx}$$

$$= \frac{1}{u} \cdot \sec^2 x$$

$$= \frac{1}{\tan x} \cdot \sec^2 x$$

$$= \frac{\sec^2 x}{\tan x}$$

B.4   ANTIDERIVATIVES

Because $(\ln x)' = \frac{1}{x}$,

$$\int \frac{1}{x} \, dx = \ln x + C \tag{B.18}$$

You can now integrate the one power of x, $x^{-1} = \frac{1}{x}$, to which the power rule did not apply. Also, by Equation B.17, you can handle any integrand that is in fractional form, and for which the numerator is (except for a multiplicative constant) the derivative of the denominator, because

$$\int \frac{1}{u}\frac{du}{dx} \, dx = \int \frac{1}{u} \, du = \ln u + C \tag{B.19}$$

Example B.4 Evaluate $\displaystyle\int \frac{x^2}{x^3 + 1} \, dx$.

Solution: Notice that the numerator is almost equal to the derivative of the denominator. Thus let $u = x^3 + 1$, so that $\dfrac{du}{dx} = 3x^2$,

$du = 3x^2\, dx$, and $x^2\, dx = \dfrac{1}{3}\, du$. With these substitutions,

$$\int \frac{x^2}{x^3 + 1}\, dx = \int \frac{1}{3} \cdot \frac{1}{u}\, du$$

$$= \frac{1}{3} \ln u$$

$$= \frac{1}{3} \ln(x^3 + 1) + C$$

Example B.5 Evaluate $\displaystyle\int \tan x\, dx$.

Solution: The trick here is to notice that if you write $\tan x = \dfrac{\sin x}{\cos x}$, then the numerator is almost the derivative of the denominator. Thus let $u = \cos x$, so that $\dfrac{du}{dx} = -\sin x$ and $\sin x\, dx = -du$. With these substitutions,

$$\int \tan x\, dx = \int \frac{\sin x}{\cos x}\, dx$$

$$= \int \frac{-1}{u}\, du$$

$$= -\ln u$$

$$= -\ln(\cos x) + C$$

From the fact that $(e^x)' = e^x$, you have that

$$\int e^x\, dx = e^x + C \tag{B.20}$$

From Equation B.15, it follows that

$$\int e^u \frac{du}{dx}\, dx = \int e^u\, du = e^u \tag{B.21}$$

So, you can handle any integrand involving an exponential, as long as the integrand is multiplied by the derivative of the term in the exponent.

Example B.6 Evaluate $\displaystyle\int x^2 e^{x^3}\, dx$.

Solution: Let $u = x^3$, so that $\dfrac{du}{dx} = 3x^2$, and $x^2\,dx = \dfrac{1}{3}\,du$. Then

$$
\begin{aligned}
\int x^2 e^{x^3}\,dx &= \int \frac{1}{3}e^u\,du \\
&= \frac{1}{3}e^u \\
&= \frac{1}{3}e^{x^3} + C
\end{aligned}
$$

Example B.7 Evaluate $\displaystyle\int [\sin(e^x)]e^x\,dx$.

Solution: For this problem, let $u = e^x$, so that $\dfrac{du}{dx} = e^x$, and $e^x\,dx = du$. Then

$$
\begin{aligned}
\int [\sin(e^x)]e^x\,dx &= \int \sin u\,du \\
&= -\cos u + C \\
&= -\cos(e^x) + C
\end{aligned}
$$

ANSWERS TO EXERCISES

Chapter 1

1. $y = \frac{5-3x}{4}$, domain $= (-\infty, \infty)$; $x = \frac{5-4y}{3}$, domain $= (-\infty, \infty)$

2. $y = \frac{1-2x}{3x}$, domain = all real numbers except for 0; $x = \frac{1}{3y+2}$, domain = all real numbers except for $\frac{-2}{3}$

3. $y = \frac{5}{4x^3-6}$, domain = all real numbers except for $\left(\frac{3}{2}\right)^{\frac{1}{3}}$; $x = \left(\frac{5+6y}{4y}\right)^{\frac{1}{3}}$, domain = all real numbers except for 0

4. a. 17   b. 31   c. $2x^2 - 1$ d. $8x^2 - 8x + 1$ e. $2a^2 - 4a + 1$
   f. $2x^2 + 4xh + 2h^2 - 4x - 4h + 1$

5. $xy^3 + 4y^2 = 16y - 5$

6.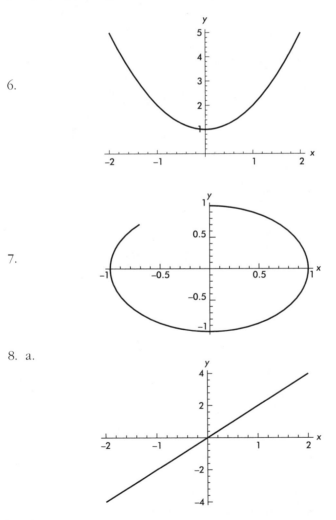

7.

8. a.

b.

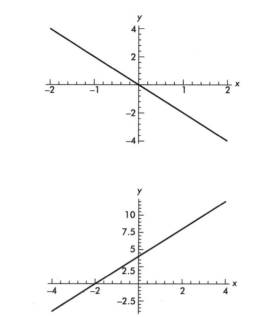

c.

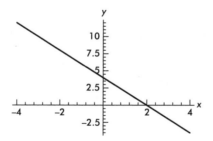

d.

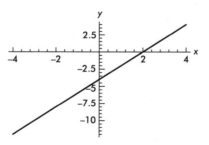

e.

f.

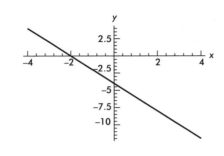

g.

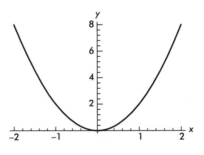

h.

i.

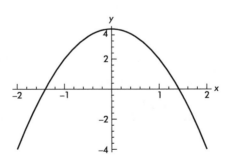

j.

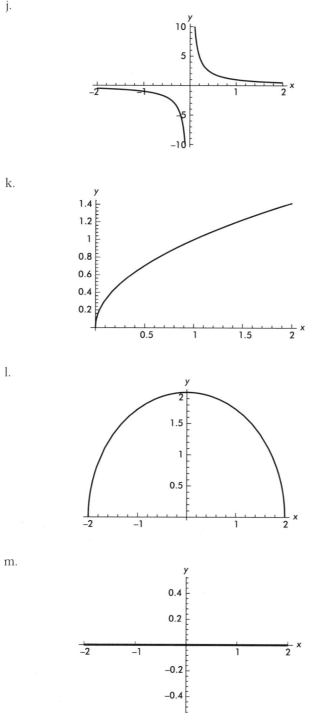

k.

l.

m.

n.

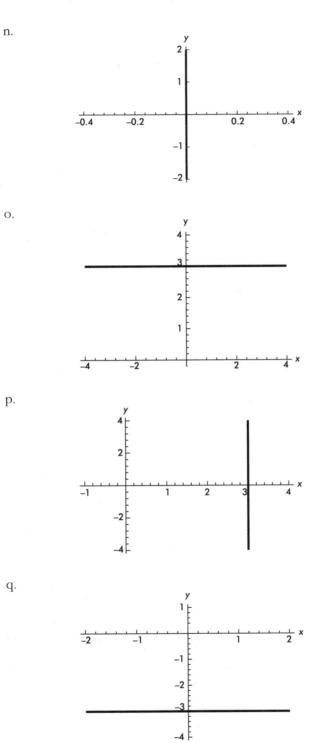

o.

p.

q.

r.

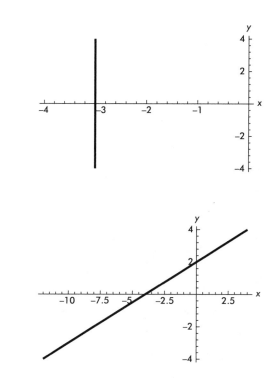

s.

Chapter 2

1. a. 7   b. 3
2. a. 5   b. 1
3. a. 4.1   b. 0.1
4. a. 4.01   b. 0.01
5. a. $4+h$ b. h
6. a. 4   b. 0
7. a. $2a+h$ b. $2a+h-4$
8. a. $2a$ b. $2a-4$
9. a. 3 seconds   b. 144 feet   c. $1\frac{1}{2}$ and $4\frac{1}{2}$ seconds
   d. -96 feet per second
10. a. $\sqrt{15}=3.87$ seconds   b. $-32\sqrt{15}=-123.84$ feet per second
    c. -16 feet per second   d. -32 feet per second
    e. $\frac{-240}{\sqrt{15}}=-62.02$ feet per second
11. a. $(2,0)$ b. $(7,25)$ c. $(2.05,0.0025)$
12. a. $(2.5,7.75)$ b. $(1.5,1.75)$ c. $x=3$

Chapter 3

1. 6
2. 0.8
3. 4
4. −12
5. 0
6. 0
7. ∞
8. ∞
9. ∞
10. does not exist
11. ∞
12. −∞
13. 6
14. 0
15. a. 2   b. 5   c. does not exist
16. a. 5   b. 4   c. does not exist
17.

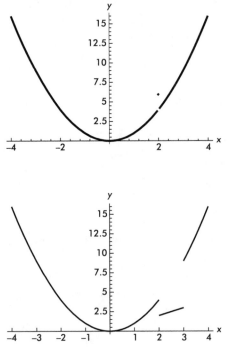

18.

19.

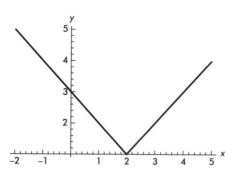

20.

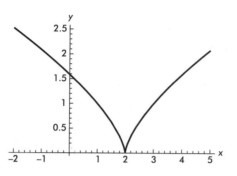

Chapter 4

1. $f'(x) = 6x^2$

2. $f'(x) = 6x + 4$

3. $f'(x) = 3x^2 - 12x$

4. $h'(t) = \frac{-3}{7t^2}$

5. $p'(x) = \frac{-25}{(7x-6)^2}$

6. $g'(t) = \frac{-2t}{(t^2+1)^2}$

7. $h'(u) = \frac{3}{2\sqrt{3u}}$

8. $p'(t) = \frac{2}{\sqrt{4t-2}}$

9. $f'(x) = \frac{x}{\sqrt{x^2+1}}$

10. $f'(x) = \dfrac{-3}{2(3x-4)^{\frac{3}{2}}}$

Chapter 5

1. $f'(x) = 4x - 3$

2. $g'(t) = 21t^2 - 8t + \frac{1}{2}$

3. $p'(s) = 18s^2 - \frac{3}{2} \cdot \frac{1}{\sqrt{s}} - \frac{4}{s^2} + \frac{22}{s^3}$

4. $h'(u) = -316u^{78}$

5. $g'(x) = 1 - \frac{1}{x^2} - \frac{2}{x^3}$

6. $f'(x) = \frac{2}{x^{\frac{1}{3}}} + \frac{2}{\sqrt{x}}$

7. $s'(t) = \frac{-21}{t^4} - \frac{2}{3t^5}$

8. $p'(u) = \frac{1}{2\sqrt{u}} + \frac{1}{2u\sqrt{u}}$

9. $f'(x) = (7x^5 - 5x^3 + 6x^2 - 4) \cdot (32x^3 - 18x)$
$\qquad + (35x^4 - 15x^2 + 12x) \cdot (8x^4 - 9x^2 + 24)$

10. $f'(t) = \sqrt{t} \cdot (2t - 1) + \frac{t^2 - t}{2\sqrt{t}}$

11. $h'(x) = \frac{-2x^2 + 2x + 1}{(x^2 - x + 1)^2}$

12. $p'(t) = \frac{3t^4 - 10t^2 - 1}{(t^2 - 1)^2}$

13. $f'(x) = \frac{(3\sqrt{x}+1)(4x-4) - (2x^2 - 4x + 5)\left(\frac{3}{2\sqrt{x}}\right)}{(3\sqrt{x}+1)^2}$

14. $h'(v) = \frac{-15}{2(5v-2)^{\frac{3}{2}}}$

15. $f'(s) = 2s \cdot (3s^2 - 6s + 9)^3 \cdot (15s^2 - 18s + 9)$

16. $p'(t) = \frac{33t^2(t^2 - 3t)^2}{(2t^2 + 5t)^4}$

17. $f'(x) = \frac{1}{\sqrt{2x}}$

18. $h'(x) = \frac{21x^2 + 8x}{2\sqrt{7x^3 + 4x^2}}$

19. $p'(t) = 68(2t)^{33}$

20. $g'(t) = \frac{12t^4 - 24t^2 + 90t + 12}{(3t^2 - 3)^2}$

21. $h'(s) = 4\left(\sqrt{5s^2 + 1} + s^3\right)^3 \left(\frac{5s}{\sqrt{5s^2+1}} + 3s^2\right)$

22. $f'(x) = \frac{4}{3}x^{\frac{1}{3}} + \frac{4}{3}\frac{1}{x^{\frac{7}{3}}}$

23. $p'(t) = (3t^2 - 1)^2(4t + 9)^4(132t^2 + 162t - 20)$

24. $h'(x) = \frac{(x^2+1)^3(2x^2 - 24x - 14)}{(2x-3)^8}$

25. $f'(x) = 3\left[3x - (x^4 + x)^5\right]^2 \left[3 - 5(x^4 + 4)^4(4x^3 + 1)\right]$

Chapter 6

1. maximum = 5, minimum = −22

2. maximum = 10, minimum = −17

3. maximum = 2, minimum = $\frac{-2}{3}$

4. maximum = 80, minimum = 16

5. maximum = $\frac{5}{4}$, minimum = −1

6. maximum = 0, minimum = −1

7. maximum = 0, minimum = $\frac{-3}{4}$

8. $c = 2$

9. $c = 1$

10. $c = \pm\sqrt{\frac{7}{3}}$

11. $c = \sqrt{\frac{7}{3}}$

12. $c = \sqrt{2}$

13.

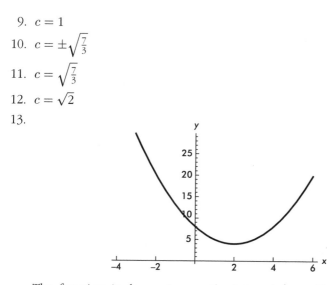

The function is decreasing on the interval $(-\infty, 2)$, increasing on the interval $(2, \infty)$, has no local maximum value, and has a local minimum at the point $(2, 4)$.

14.

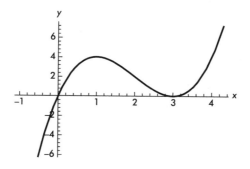

The function is increasing on the intervals $(-\infty, 1)$ and $(3, \infty)$, is decreasing on the interval $(1, 3)$, has a local maximum at the point $(1, 4)$, and has a local minimum at the point $(3, 0)$.

15.

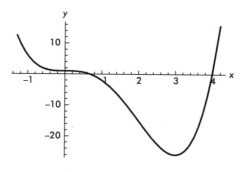

The function is decreasing on the interval $(-\infty, 3)$, is increasing on the interval $(3, \infty)$, has no local maximum, and has a local minimum at the point $(3, -26)$. NOTE: The first derivative tells you that the function is decreasing on $(-\infty, 3)$, but gives no additional information about the shape of the curve on this interval. That requires the second derivative.

16.

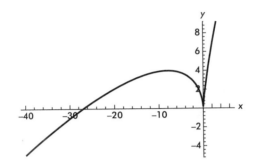

The function is increasing on the intervals $(-\infty, -8)$ and $(0, \infty)$, decreasing on the interval $(-8, 0)$, has a local maximum at the point $(-8, 4)$, and has a local minimum at the point $(0, 0)$.

Chapter 7

1. 200 feet × 100 feet, with the longer side parallel to the river

2. 50 feet × 50 feet

3. $4\frac{2}{3}$ feet × $4\frac{2}{3}$ feet × 7 feet

4. approximately 176.5 feet × 214 feet, with the shorter side parallel to the street

5. 16 square inches

6. a^2 square inches

7. radius = $\left(\frac{150}{\pi}\right)^{\frac{1}{3}}$ inches, height = $2\left(\frac{150}{\pi}\right)^{\frac{1}{3}}$ inches

8. radius = height = $\left(\frac{300}{\pi}\right)^{\frac{1}{3}}$

9. volume = $\frac{25}{3\sqrt{30\pi}}$ cubic inches

10. $x = 33\frac{1}{3}, y = 16\frac{2}{3}$

11. $x = 0, y = 50$, or $x = 50, y = 0$

12. $x = y = 10$

13. There is no answer. The sum can be made arbitrarily large by choosing one of the numbers close enough to 0.

14. approximately 74.6 miles per hour

15. $4\sqrt{2}$ inches per second

16. 24 feet per second

17. $6\sqrt{10}$ feet per second

18. $\frac{40\sqrt{2}}{3}$ feet per second

19. $\frac{27}{20\pi}$ feet per second

20. $\frac{3}{20\pi}$ feet per second

21. π cubic inches per minute

22. $\frac{\pi}{2}\left(\frac{16}{\pi}\right)^{\frac{2}{3}}$ cubic inches per minute

23. $\frac{2}{(4\pi)^{\frac{1}{3}}3^{\frac{2}{3}}}$ feet per minute

24. 4704π cubic feet per minute

25. 48π square feet per second

26. $40\sqrt{\pi}$ square feet per second

27. 1.5

28. 2

Chapter 8

1. 9.915

2. 13.3935

3. 17.511

4. 10.625

5. 13.4375

6. 16.625

7. $x_i = \frac{i}{2}$; $\Sigma_{i=1}^{6}\left[\left(\frac{i}{2}\right)^2 + \frac{i}{2}\right] \cdot \frac{1}{2}$

8. $\Sigma_{i=1}^{100}\left[\left(\frac{3i}{100}\right)^2 + \frac{3i}{100}\right] \cdot \frac{3}{100}$

9. $\Sigma_{i=1}^{n}\left[\left(\frac{3i}{n}\right)^2 + \frac{3i}{n}\right] \cdot \frac{3}{n}$

10. $\lim_{n\to\infty}\Sigma_{i=1}^{n}\left[\left(\frac{3i}{n}\right)^2 + \frac{3i}{n}\right] \cdot \frac{3}{n}$

11. $\int_0^3 (x^2 + x)\, dx$

12. 33

13. 166

14. $\Sigma_{i=1}^{97} 3i^2$

15. $\Sigma_{i=1}^{10} 2i$

16. $\Sigma_{i=1}^{19}(2i - 1)$

17. $\lim_{n\to\infty}\Sigma_{i=1}^{n}\left[1 + \left(\frac{2i}{n}\right)^2\right] \cdot \frac{2}{n}$

18. $\lim_{n\to\infty}\Sigma_{i=1}^{n}\left[2(1 + \frac{2i}{n})\right] \cdot \frac{2}{n}$

19. $\lim_{n\to\infty}\Sigma_{i=1}^{n}\left[(-2 + \frac{5i}{n})^2 + 4\right] \cdot \frac{5}{n}$

20. 198 feet

Chapter 9

1. 164

2. 140

3. 3,493,503

4. 18,354

5. $4n - \frac{3n(n+1)}{2}$

6. $\frac{n(n+1)(2n+1)}{6} + \frac{3n(n+1)}{2} - 2n$

7. ∞

8. $\frac{3}{2}$

9. 0

10. $\frac{45}{2}$

11. 20

12. $\frac{20}{3}$

13. 64 feet

14. $\frac{-8}{3}$

Chapter 10

1. $F(x) = x^3 - 8, F'(x) = 3x^2$

2. $F(x) = \frac{2}{3}x^{\frac{3}{2}} + x - \frac{5}{3}, F'(x) = x^{\frac{1}{2}} + 1$

3. $F'(x) = \frac{1}{x}$

4. $F'(x) = (x^3 - 1)^{11}$

5. $\frac{2}{7}x^7 - \frac{3}{5}x^{10} + C$

6. $(2t - 1)^{\frac{3}{2}} + C$

7. $\frac{-1}{6}\left(\frac{1}{6x+4}\right) + C$

8. $\frac{1}{3}x^3 + \frac{1}{2}x^2 - \frac{1}{x} + C$

9. $\frac{1}{2}s^4 - s^3 + 2s^2 - 7s + C$

10. $2\sqrt{y - 4} + C$

11. $\frac{-4}{3x^{\frac{3}{2}}} + C$

12. $\frac{1}{4}(2x^4 + 1)^{\frac{3}{2}} + C$

13. $(t^2 + 4)^{\frac{1}{2}} + C$

14. $\frac{-1}{18(s^3 - s)^6} + C$

15. $\frac{1}{3}(\sqrt{x} + 1)^6 + C$

16. $\frac{9}{2}$

17. $\frac{3}{16}$

18. 9687.5

19. $\frac{74}{3}$

20. -2

Chapter 11

1. $\frac{38}{3}$
2. $\frac{4}{3}$
3. 1
4. $\frac{2}{3}$
5. $\frac{8}{3}$
6. $\frac{4}{3}$
7. $\frac{2}{3}$
8. $\frac{8}{3}$
9. $\frac{9}{2}$
10. $\frac{8}{3}$
11. $\frac{128\pi}{7}$
12. $\frac{104\pi}{5}$
13. $\frac{64\pi}{5}$
14. $\frac{56\pi}{5}$
15. $\frac{128\pi}{7}$
16. $\frac{104\pi}{5}$
17. $\frac{64\pi}{5}$
18. $\frac{56\pi}{5}$
19. $\frac{48\pi}{5}$
20. $\frac{24\pi}{5}$
21. $\frac{48\pi}{5}$
22. $\frac{24\pi}{5}$
23. $\frac{16\pi}{3}$
24. $\frac{16\pi}{3}$
25. $\frac{2}{3}\pi a^3$
26. $\frac{2}{3}\pi a^3$

Chapter 12

1. $y = 2\sqrt{t} - 1$
2. $x = \frac{1}{4}t^4 - \frac{3}{2}t^2 + 1$
3. $y = 2x^2 - \frac{1}{x} - \frac{15}{2}$
4. a. $v = -32t - 64$, $s = -16t^2 - 64t + 192$ b. 2 seconds
5. a. $v = -32t$, $s = -16t^2 + 192$ b. $2\sqrt{3} = 3.46$ seconds
6. 336 feet
7. 100 feet per second
8. 112 feet per second

GLOSSARY

Acceleration. The rate of change of velocity with respect to time.

Antiderivative. If $F(x)$ is a function whose derivative equals $f(x)$, then $F(x)$ is an antiderivative of $f(x)$.

Approximate sum. Let $f(x)$ be a function on the interval $[a, b]$, $a = x_0 < x_1 < \ldots < x_n = b$ a partition of $[a, b]$, $\Delta x_i = x_i - x_{i-1}$, and p_i a point in $[x_{i-1}, x_i]$. An approximate sum is a sum of the form $\Sigma_{i=1}^{n} f(p_i) \Delta x_i$. Approximate sums can be used to estimate many quantities. For example, if $f(x) \geq 0$ on $[a, b]$, then the approximate sum estimates the area under the graph of $f(x)$ and over the x-axis, between $x = a$ and $x = b$. The definite integral of f from a to b, denoted $\int_a^b f(x)\, dx$, is a limit, in a suitable sense, of approximate sums.

Average rate of change. If y and x are two quantities, the average rate of change of y with respect to x is a change in y divided by a corresponding change in x.

Calculus. That area of mathematics that employs the fundamental idea of *limits* for the study and analysis of quantities that are changing. The main topics in first-year calculus are the definition, computation, and application of derivatives and integrals (antiderivatives).

Cartesian coordinates. The system, named after René Descartes, for setting up a correspondence between points in the plane and (ordered) pairs of real numbers.

Chain rule. The rule for finding the derivative of a composite func-tion. If $F(x) = g(f(x))$, then $F'(x) = g'(f(x))f'(x)$.

Continuous. A function $f(x)$ is continuous at a point $x = a$ if the values $f(x)$ can be made as close as you wish to the value $f(a)$ by choosing x close enough to (although not equal to) a. A function is continuous on an interval if it is continuous at every point of the interval.

Critical point. A function $f(x)$ has a critical point at $x = c$ if f is defined at c, and either $f'(c) = 0$ or $f'(c)$ does not exist.

Cross-sectional areas, method of. The method of finding the volume of a solid by integrating the cross-sectional areas perpendicular to some axis. If the solid is a solid of revolution, the cross-sectional areas are swept out by lines, in the region being revolved, that are perpendicular to the axis of revolution.

Cylindrical shells, method of. The method of finding the volume of a solid of revolution by integrating the lateral surface area of cylinders that are swept out by lines, in the region being revolved, that are parallel to the axis of revolution.

Definite integral. If $f(x)$ is a function, the definite integral of f from a to b, written as $\int_a^b f(x)\, dx$, is the value $F(b) - F(a)$, where $F(x)$ is any anti-derivative of $f(x)$. The definition of the definite integral is in terms of lim-its of Riemann sums, or approximate sums. The fact that definite integrals can be computed by antiderivatives is the main content of the Fundamen-tal Theorem of Calculus. Definite integrals can be used to compute areas, volumes, and much more.

Derivative. For a function $y = f(x)$, the derivative of f with respect to x, denoted by $f'(x)$ or $\dfrac{dy}{dx}$, is the instantaneous rate of change of f with respect to x. It is defined as $\lim_{h \to 0} \dfrac{f(x + h) - f(x)}{h}$, a limit of average rates of change of f with respect to x. A geometric interpretation of $f'(x)$ is as the slope of the tangent line to the graph of f, at the point $(x, f(x))$ on the graph.

Differentiable. A function f is differentiable at the point x if the limit involved in the definition of the derivative $f'(x)$ exists.

Equation. A formula stating that two mathematical expressions have the same value.

Extreme value. A maximum or minimum value of a function on an interval.

Function. A formula, rule, or procedure for taking a set of numbers and assigning to each number in the set precisely *one* value. Actually, functions can be defined on more general sets than sets of numbers, and their values can also be in more general sets than sets of numbers.

Fundamental Theorem of Calculus. The theorem that states that if $f(x)$ is a continuous function on the interval $[a, b]$, then the function $A(x) = \int_a^x f(t)\,dt$ is differentiable, with $A'(x) = f(x)$. It follows that for *any* antiderivative $F(x)$ of $f(x)$, $\int_a^b f(x)\,dx = F(b) - F(a)$.

Graph. For an equation involving two variables x and y, the graph of the equation consists of all points (a, b) in the plane such that when the value a is substituted for x and b for y, the equation holds true.

Implicit derivative. The derivative of a function that is not defined explicitly in terms of the variable.

Indefinite integral. The general antiderivative of a function. If $f(x)$ is a function and $F(x)$ is one antiderivative of $f(x)$, then the indefinite integral of $f(x)$, denoted by $\int f(x)\,dx$, is of the form $F(x) + C$, for C is an arbitrary constant.

Initial value problem. A problem in which you are given a relationship between a variable, an unknown function of that variable, and the derivative of the function, as well as the value of the function at some particular point, and are asked to find the unknown function.

Instantaneous rate of change. If y and x are two quantities, the instantaneous rate of change of y with respect to x is the limit of average rates of change of y with respect to x, as the change in x gets closer and closer to 0. If $y = f(x)$, the instantaneous rate of change of y with respect to x is the derivative $f'(x)$.

Integral. Either a definite or indefinite integral.

Integrand. The function being integrated.

Integration by substitution. A technique for finding the antiderivative of certain functions that is based upon the chain rule for derivatives.

Limit. A value that a collection of terms is getting arbitrarily close to. If

$f(x)$ is a function, then the limit of $f(x)$, as x approaches a, equals L, written $\lim_{x \to a} f(x) = L$, if the values of $f(x)$ can be made arbitrarily close to the number L simply by choosing x close enough to (but not equal to) a. If $\lim_{x \to a} f(x) = f(a)$, then f is continuous at $x = a$.

Limits of integration. The endpoints of the interval over which a definite integral is to be computed. In the definite integral $\int_a^b f(x)\,dx$, a is called the *lower limit of integration* and b is called the *upper limit of integration*.

Maximum value. The largest value that a function has on an interval.

Mean value theorem. The theorem that states that if a function $f(x)$ is continuous on a closed and bounded interval $[a, b]$, and differentiable on the interval (a, b), then there is at least one point c, with $a < c < b$, such that $f'(c) = \dfrac{f(b) - f(a)}{b - a}$.

Minimum value. The smallest value that a function has on an interval.

Partition. A division of an interval into subintervals.

Power rule for derivatives. The rule that says that for $f(x) = x^n$, then $f'(x) = nx^{n-1}$.

Power rule for integrals. The rule that says that for $n \neq -1$, $\int x^n\,dx = \dfrac{1}{n+1}x^{n+1} + C$.

Product rule. The rule for finding the derivative of a product of two functions. If $F(x) = f(x)g(x)$, then $F'(x) = f(x)g'(x) + f'(x)g(x)$.

Quotient rule. The rule for finding the derivative of a quotient of two functions. If $F(x) = \dfrac{f(x)}{g(x)}$, then $F'(x) = \dfrac{g(x)f'(x) - f(x)g'(x)}{[g(x)]^2}$.

Regular partition. A division of an interval into subintervals of equal length.

Riemann sum. The same as an approximate sum. Named after Georg F. B. Riemann (1826–1866), a German mathematician.

Rolle's theorem. A special case of the mean value theorem, which applies in the case that $f(a) = f(b) = 0$.

Speed. The absolute value of velocity.

Slope. If l is a non-vertical line, and (x_1, y_1) and (x_2, y_2) are any two distinct points on l, then the slope of the line l is the ratio $\dfrac{y_2 - y_1}{x_2 - x_1}$. The slope of a vertical line equals ∞. If $f(x)$ is a function, then the slope of the tangent line to the graph of f, at the point $(a, f(a))$ on the graph, is given by the derivative $f'(a)$.

Solid of revolution. The three-dimensional volume formed when a region or area in the plane is revolved about a line.

Tangent line. If $y = f(x)$ is a function, then the tangent line to the graph of f, at the point $(a, f(a))$ on the graph, is the straight line that passes through the point $(a, f(a))$, and that has slope $f'(a)$.

Velocity. The rate of change of position with respect to time.

Vertical line test. A graph is the graph of a *function* if every vertical line crosses the graph at most once.

INDEX